EXPLORING CONSTRUCTION

By

Richard M. Henak
Professor of Industry and Technology Emeritus
Ball State University
Muncie, Indiana

Publisher
The Goodheart-Willcox Company, Inc.
Tinley Park, Illinois

ABOUT THE AUTHOR

In addition to Exploring Construction, Richard M. Henak is also the coauthor of *Exploring Production*. Dick has a Bachelor of Arts degree from the University of Northern Iowa, a Master of Arts degree from Ball State University, and a Doctorate of Education degree from the University of Illinois. Dick is a retired professor of Industry and Technology at Ball State University. Before retirement, Dick was also actively involved in the construction trade and construction education, and was self-employed in the construction field.

Copyright 2000

by

THE GOODHEART-WILLCOX COMPANY, INC.

Previous Editions Copyright 1993, 1985

Library of Congress Catalog Card Number 99-17782
International Standard Book Number 1-56637-681-5

2 3 4 5 6 7 8 9 10 00 03 02 01 00

Cover Photo: Nawrocki Stock Photo Inc.

Library of Congress Cataloging-in-Publication Data

Henak, Richard M.
    Exploring construction / Richard M. Henak.

        p. cm.
    Includes index.
    ISBN 1-56637-681-5
    1. Building. 2. Building trades -- Vocational guidance.
    I. Title.
    TH146.H45   1999
    624--dc21   99-17782
                                                     CIP

INTRODUCTION

EXPLORING CONSTRUCTION describes the construction industry and methods used in construction. You will be introduced to ways construction technology provides structures and projects like highways, dams, and buildings. You will learn what work construction firms do and what construction trades perform this work.

EXPLORING CONSTRUCTION teaches you about three major areas involved with construction: planning the project, building the project, and servicing the project. You will learn about construction management, construction technology, job scheduling, hiring, contracting, contracts, and final payments. You will see how projects are transferred to an owner when construction is complete, and how projects are operated.

EXPLORING CONSTRUCTION will help you understand how a building is constructed from the beginning of the foundation to the completion of the roof. It also discusses the installation of the mechanical systems, including the electrical and plumbing systems. Work on many other mechanical systems is discussed. Finally, you will learn about methods of doing finishing work that will complete the structure.

You will explore how construction affects the world about you and how construction technology affects decisions you make. You will look at ways construction technology can improve life in the future.

You will have the opportunity to work on building projects with your class. This will help you find out what kinds of work you like to do. You will also have the opportunity to read about exciting careers in construction.

Richard M. Henak

3

CONTENTS

Section 1
STUDYING CONSTRUCTION TECHNOLOGY

Chapter 1

INTRODUCTION TO CONSTRUCTION TECHNOLOGY

After studying this chapter, you will be able to:
☐ Describe living conditions before the use of construction technology.
☐ List problems that were solved by construction technology.
☐ Explain why progress is possible for humans but not for animals.

We can learn about a subject in more than one way. Let us use cars as an example. We can learn about the cars of the past. We can describe the kinds of cars. We can learn about how they were built and about their future.

We will study construction technology in all of the above ways. We will learn about:
• Construction of the past.
• The kinds of construction projects.
• How construction projects are built.
• Construction in the future.

This chapter will discuss construction in the past. It will give you a background for construction today. Other chapters will be about current construction practices and how to begin a project. Toward the end of this book we will look at construction and the future.

CONSTRUCTION IN THE PAST

Construction technology is the study of how houses, commercial buildings, and roads are built. A *construction project* is built where it is used. Its foundation depends on the earth to hold it. The project may be built above or below the surface of the ground. You can see many construction projects in Fig. 1-1. There are many you cannot see, too. They are the ones beneath the surface of the ground. Water, sewer, and natural gas lines are examples of these projects.

Fig. 1-1. If people could not build construction projects, what would you see in this picture? (Mitchell/Giurgola Architects)

We will briefly trace the growth of construction technology. We will look at housing, sanitary systems, solid waste systems, roadways, industrial construction, and energy systems.

Housing

People live in houses and use them for *shelter,* protection from the elements. Early people would gather plants, and they would hunt and fish for meat. They did not have houses as we know them. See Fig. 1-2. They had to keep moving because animals used for food would migrate, and food and plants were spread over a wide area. Some of the people would sleep overnight in crude nests in the trees. It was safer in the trees than on the ground. Others stayed in caves or under shelters made of tree branches, grasses, and leaves.

Then people learned to grow plants and raise animals for food. They started to build huts. The huts

Fig. 1-2. When people had to keep moving from place to place, they could not build permanent structures.

houses were found. People learned to use sun-baked clay bricks and to cut stone.

The Romans began using central heat. Heated air from ovens was moved through tubes to heat the floors of rooms.

The people who first came to North America from other lands brought their ways of building houses with them. They used many of their own methods. But, the demands and materials of the new land changed some of their ideas.

Before the 17th century, craftspersons would design the houses. Bricklayers, carpenters, and others did most of the design work. Starting in the 1700s, people called architects began to design the houses.

Houses today are mostly single family dwellings, Fig 1-4. The number of people has grown and the cost of land has increased. As time goes by, less land is used for each structure. Different kinds of structures are designed to look similar and share common areas, Fig. 1-5. Some homes are built together with no land between them, Fig. 1-6. Fig. 1-7 shows an apartment complex. An apartment building can save on materials and heating costs. A high-rise apartment building made of reinforced concrete is shown in Fig. 1-8. These structures are built in and around cities.

were made of stone, snow, clay, leaves, hides, and wood. Huts were sometimes built over a hole to make more head room.

People began to join into groups. They formed villages. Fences were built to hold in animals and to hold out enemies, Fig. 1-3. New ways to build

Fig. 1-3. What construction projects do you see in this picture?

Fig. 1-4. This is a single family dwelling. Homes like these are being built smaller and on less land than before. (Cedar Shake and Shingle Bureau)

Fig. 1-7. Construction materials and space can be saved by combining dwellings into one structure, such as this apartment complex.
(California Redwood Assoc.)

Fig. 1-5. Two six-million-gallon water storage reservoirs share a "corporate yard" with two buildings. (DYK Incorporated)

Fig. 1-8. This structure has homes in it. They are called condominiums. In this building, each home is owned by the people who live in it. (Dover Elevators)

Sanitary Systems

Sanitary systems are used to provide clean water and to get rid of household and human waste.

Water supplies

The first record of a water supply system dates back 5000 years. The system was in the city of Nippur in Sumeria. Water was drawn from wells and hand-carried to the homes.

Water treatment is used to make water clean enough to drink. Water treatment was first used around 2000 B.C. The methods used were much like those used today. They would boil the water, filter it through charcoal, and expose it to sunlight.

Fig. 1-6. Notice there is no land between these homes. They use less land and are easier to heat and cool. Why? (California Redwood Assoc.)

Today, making water clean enough to drink is a big problem. Some reasons are:
- We need more water than ever before.
- There is much more waste in the water.
- Much of today's waste presents a greater danger to our health.
- Some waste in the water now is much harder to remove.

Aqueducts were used to carry water into the cities. See Fig. 1-9. Romans started to construct their aqueducts about 312 B.C. Aqueducts, like rivers, use gravity to move water.

The Greeks invented pumps. The Romans were the first to use them to draw water for a city. Later, the English used pumps to move water through to their homes. Hollow logs were used for mains (large water pipes). The pumps were first run by a water-wheel. Later, steam engines were used. We now use electric motors, engines, and turbines to run our pumps. In the southwestern United States, water is pumped through pipes for nearly 250 miles. See Fig. 1-10.

Sewers

In early times, it was easy to get rid of human and household wastes. There was plenty of space and fewer people. As cities got larger, the need for a sewer system grew. In Rome, sewers were used to get rid of surface water. The Roman storm sewer system was built over 2500 years ago. It is still in service today.

Human and household wastes were thrown in the streets. People expected the rain to wash it away. The filth accumulated (collected) and bacte-

Fig. 1-10. Water is brought to cities in pipes like this. (Wire Reinforcement Institute)

ria grew in it. Laws and the invention of the flush toilet caused a great change to occur. New laws would no longer let people put their waste in the street; it had to be put in sewers. Those two events solved one health problem, but led to the first problem of water pollution in Great Britain.

In 1850, the relationship between a water supply, sewers, and health was seen. Massachusetts was the first state to form a board of health. This board had a doctor, lawyer, chemist, civil engineer, and three others as members.

Today, sewage systems do a better job of cleaning wastewater. We now separate storm sewers and sanitary sewers. The storm water is not treated. It takes a direct route to the waterways. The sanitary sewers lead to the wastewater treatment plant. See Fig. 1-11.

Solid Waste

When this country was young, people spent their time trying to survive. They did not have many things. Little was thrown away. Things that were thrown away were usually substances that decayed naturally (called organic wastes). Getting rid of solid waste was not a problem.

We had plenty of resources. Problems began when cheap energy and machines were used to produce more things. New things replaced old things at a faster rate. We began to package our products.

Fig. 1-9. This is a drawing of an aqueduct that is still used today. It is in Rome.

Fig. 1-11. All cities and most towns have a wastewater treatment plant that removes contaminants. (DYK Incorporated)

More products and packages were thrown away by more people. This was added to organic wastes. We soon had a solid waste problem.

A *solid waste system* had to be developed to handle all the solid waste produced.

Open dumping and burning of solid waste (which includes construction wastes) was used first. Then, waste was buried in holes. The practice of using sanitary landfills started during World War II. In a *sanitary landfill,* items are covered soon after dumping.

Where there is a lot of people and little land, new practices are being used. Burnable solid waste can be baled so it takes up less space. It can be burned under certain conditions. A practice called resource recovery is being used. Plants are built where the trash is burned to make steam. The steam is used for heating buildings or to make electricity. Some plants recycle the metals and aggregate (glass and rock). You can see how these plants work in Fig. 1-12.

Sometimes organic wastes will not burn but will decay. They can be used for energy by putting them into a digester. The *digester* uses bacteria to convert some of the waste into methane gas. The gas can

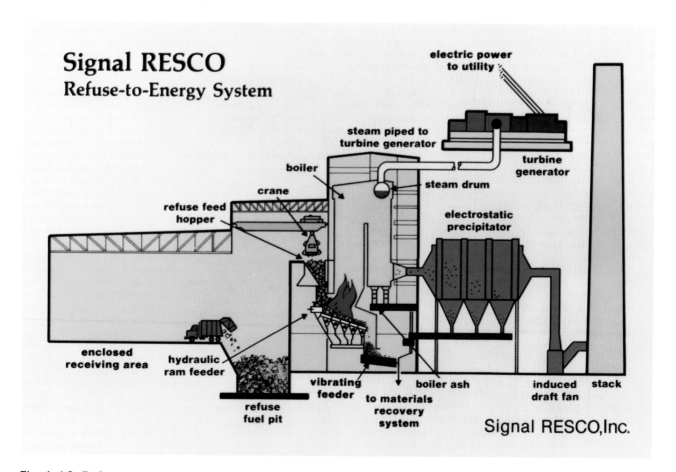

Fig. 1-12. Refuse-to-energy systems are being built where there is enough trash. This one is on the east coast. (Signal RESCO, Inc.)

then be used in internal combustion engines and for heating. Many commercial systems have been built to do this.

Roadways

Roadways are flat structures. They are used to support vehicles (cars, trains, and airplanes). Vehicles move people and things.

Roads

The first roads were built for a single purpose. People in Egypt built a single-purpose road. Their road was used to move materials to construct the Great Pyramids.

The city of Babylon was paved as early as 2000 B.C. The Roman Empire, at its peak, had 50,000 miles of roads. These roads were built for their armies. Roads used by armies are called military roads.

Roads in this country started as animal paths. The route was not always the shortest. It was, however, easy to travel. The paths went around things that were in the way. The paths were made wider so carts and wagons could use them. Rock and gravel were used to make the surface hard and level. Bridges were built. Gradually the paths became roads. Concrete began to be used.

We now know how to build roads over, under, and through things that are in the way. The new roads save time and they make going places easier. We use the tops of hills to fill valleys. Bridges, large and small, are built over water, canyons, and other roadways, Fig. 1-13.

Railroads

Four things make up a *railroad.* The first is a pulling unit. It is called a locomotive. The second is single cars hooked to each other to form a train. The track on which the train runs is the third. Finally, to be a railroad the train must provide a common carrier service. This means to haul freight or people. See Fig. 1-14.

Fig. 1-14. This is a railroad. There is a power unit, cars linked together, track, and freight. (Association of American Railroads)

Fig. 1-13. These bridges make it safer and faster to travel. Can you explain why? (HNTB Engineers)

The first trains were used in Wales for mining. The idea of long-range movement of goods by train in America was started in the summer of 1830. The Baltimore and Ohio Railroad used horses for power. In December of 1830, the South Carolina Railroad joined all four parts of a railroad. They used a locomotive to pull a train on a track to haul freight for others.

Roadway builders are now able to build better roadways. Cars and trains travel more safely and with more ease than ever before. Trains are able to quickly move large loads over a long distance.

Industrial Construction

Early people made what they needed themselves. When they began to join into groups they started to barter. A person with extra arrows would trade them for another person's extra fur.

In time, people began to specialize. They would make many items they knew how to make or liked to make. Other items that a person needed were obtained from people who specialized in making those items. People who made the items in their homes ran an enterprise called a cottage industry.

The number of people and their needs grew. Methods used to produce products got better. Those who ran cottage industries began to pay others to help with their work. Their homes became too small. This centralized production required *industrial construction,* or the construction of large factory buildings.

The use of railroads and boats grew. Materials were brought in and products were shipped out to distant markets. The factories grew even larger. They had to produce more products. The size of the machines needed to produce products got larger, too.

Some products could only be made in huge plants. Petroleum (oil) products, metals, paper, and cars are made in large industrial complexes. They often cover hundreds of acres of land. Fig. 1-15 shows one example.

Construction of Energy Systems

The number of people grew. Their needs and wants grew, too. Energy was needed to support this growth. We needed energy transported all over the country.

Wood was the first source of energy used, but coal was the first energy source needing nationwide

Fig. 1-15. Products are made from crude oil at this industrial complex called a refinery. (Mobil Oil Corp.)

transportation. Railroads were expanded to take coal to most parts of the country.

Then oil was found. Although trains could haul oil, there were problems. To see the problem, compare coal transportation with oil transportation. Railroad owners could afford to build track to coal mines: volume was high and it was in one place. They could not afford to connect the oil wells with rail lines. The oil wells were spread out.

Pipelines, like the one in Fig. 1-16, were a better answer. Small lines ran between wells to larger lines. The larger lines took the oil to plants where it was made into products people could use. Other pipes are used to move the products to where they are sold. These pipelines use thousands of miles of pipe.

Ground-up coal can be mixed with water to make a slurry. This mushy material is pumped through pipes, too.

Electricity cannot be moved in a truck or a train. Large cables on poles and towers are used to transmit electricity. Electricity from the power plant, Fig. 1-17, is sent to substations. From the substa-

Fig. 1-17. Here is a power plant and transmission lines. These lines go to substations. Then they take the power to where it is used.
(Southern Company Services)

Fig. 1-16. Pipeline construction. Earthwork is a part of most construction projects.
(Association of Oil Pipelines)

tions it goes to homes, factories, and public buildings. This industry started just over 100 years ago.

Energy for Construction

Energy is needed to operate the machines used to build a structure, roadway, or other project. For example, machines are used to do earthwork, haul and place concrete, and lift steel beams into the correct position.

Energy is also needed to produce construction materials. It takes a huge amount of energy to make the cement used in concrete. The energy to make concrete accounts for 25 percent of all the energy used in manufacturing, construction, transportation, and communication.

Energy is an input

Energy is used in almost all processes, natural or created by man. Energy is said to be an input to these processes. Some related inputs required for a manufacturing business or a construction enterprise are: capital (money), labor, and equipment. For construction, labor is the largest input.

SOCIAL IMPACT OF CONSTRUCTION TECHNOLOGY

Without a knowledge of how to build things, little progress occurs. For example, wild horses have made very little progress over the years. The changes that did occur were the result of people (not horses). They still run wild in parts of our country.

People and horses are not the same: we are able to learn to a much greater degree. We have learned how to build structures that serve our needs. As a result, our culture has grown. Horses are unable to build structures. Therefore, they have not progressed in a similar fashion.

Effect of Construction Technology and Other Technologies

Our culture is growing at an increasing rate. We need and want more products. We learn how to produce what we need and want. But, we are at a dangerous time. The things we produce are causing some problems. The methods we use to produce them are also causing problems. We are using our sources of energy at a very fast rate. Our air, water, and land are being polluted. We must make good decisions today. If we do, we will have a good place to live in the future.

SUMMARY

When people needed to keep moving to find food, they could not build permanent structures. When they grew their own food, they could live in villages. Water supplies were one of the first construction projects. A sewer system (a type of sanitary system), carried wastes away more safely than earlier methods.

Roadways were first built with rock and gravel, and later with concrete. Industrial construction includes factory buildings. Factories were built as work became more specialized than that done in cottage industries. Pipelines were built as oil became a major source of energy.

Humans have progressed because of construction technology. We are different from animals in this way.

KEY WORDS

All of the following words have been used in this chapter. Do you know their meaning?

Construction project
Construction technology
Digester
Industrial construction
Railroad
Roadway
Sanitary landfill
Sanitary system
Shelter
Solid waste system

TEST YOUR KNOWLEDGE

Write your answers on a separate sheet of paper. Do not write in this book

1. List some examples of underground construction projects.
2. Which of the following did the Romans use to build central heating systems?
   a. Ceiling tubes.
   b. Floor tubes.
   c. Boilers.
   d. Ovens.
   e. Fireplaces.
3. To conserve land, several homes can _____ a court area.
4. Aqueducts use _____ to move water through them.

5. In modern construction, _____ sewers are built separate from waste lines.
6. What is used to make a road hard and level?
7. Waste from living and from construction is put in a sanitary _____.
8. A common building material for high-rise apartments is _____ concrete.
9. What cultural development makes construction of large buildings necessary?
10. Without a knowledge of how to build things, little _____ occurs.

ACTIVITIES

1. Trace the development of one kind of construction. You decide on the kind. Make a report in any way you choose. (Use a videotape recorder, slides, audiotape recording of an interview, digital camera, etc.)
2. Go to a library and see if they have models of constructed projects. Sketch and describe any that interest you.
3. Look around the school or where you live. See if you can find a part of a sanitary, solid waste, roadway, energy, or industrial project. Describe it.

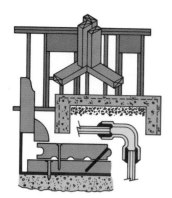

Chapter 2

WHAT IS CONSTRUCTION TECHNOLOGY?

After studying this chapter, you will be able to:
☐ Recognize items produced by the construction industry.
☐ Describe some natural materials and describe some assemblies.
☐ Compare building construction with heavy engineering construction.
☐ Give reasons for choosing careers in planning, building, or servicing.

Suppose it is raining outside, but you are warm and dry in your classroom. The structure of the building keeps the weather out. Also recall that the bus or car that brought you to school traveled over a street and perhaps a bridge. The roadway was level and smooth. These good things are the result of using construction technology to build projects.

CONSTRUCTION TECHNOLOGY

People use resources to produce the products and structures they need. Manufactured products are made in a factory. They are then taken to the point of their use. The chair in Fig. 2-1 is made in Zeeland, Michigan. It is used all over the world. The materials, machines, and workers are at the factory where the chair is made.

Construction projects are not like the chair. They are built where they are to be used. Pipelines, skyscrapers, and most homes are fixed to the building site. Unlike manufacturing resources, resources used in construction are brought to the site where the project is built.

RESOURCES

The *resources* used in construction include materials, equipment, workers, methods, and manage-

Fig. 2-1. This furniture was built in Zeeland, Michigan but is used in many parts of the world. (Herman Miller Inc.)

ment. See Fig. 2-2. Using these resources efficiently is construction technology. This idea is shown in Fig. 2-3. You can define construction technology this way. *Construction technology is using materi-*

Fig. 2-2. How many resources can you find in this picture? (Caterpillar Inc.)

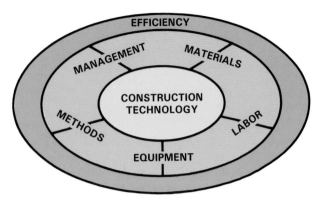

Fig. 2-3. Relationship of resources required for construction technology.

als, labor, equipment, methods, and management resources efficiently to produce a structure at a site.

Materials

Material resources may be natural materials, manufactured materials, assemblies, or finished products. The soil taken from the top of a hill and used in a valley to make a level roadbed is a *natural material*. Steel rod used to make a concrete highway strong is a *manufactured material*. A manufactured material, shown in Fig. 2-4, is a girder for a bridge. Signs for a new highway are assemblies. *Assemblies* are items that need only minor work at the site. A transformer for a power plant is a finished product. *Finished products* only need to be set in place and connected into the system.

Equipment

The second resource used in construction is *equipment.* Most new construction uses equipment to move soil. Even small projects require equipment. A rake and a spade are needed to plant shrubs around a new home. A crane barge is used to construct projects in water. It is a large and complex piece of equipment. See Fig. 2-5.

Workers

The third resource is workers. *Workers* are people who run equipment and use the materials. When building large projects in remote areas, workers live

Fig. 2-4. These assemblies are made by a steel fabricator and are assembled on the site. (Maine Department of Transportation)

Fig. 2-5. Crane barges are used to build structures over water. (The Manitowoc Company, Inc.)

Fig. 2-6. Managers plan, organize, direct, and control the work. (Southern Company Services)

in temporary housing during the week. They may travel to their homes on weekends.

Methods

Methods are needed to build projects. *Construction methods* include the use of tools and techniques to build something on a site. You will learn about tools and techniques of building. The process of building projects is in the last part of this chapter.

Management

Management is the people resource that gives leadership, Fig. 2-6. Managers usually do not build the project with their own hands. They get things done through other people. Managers plan the way to build the project. They schedule the construction effort. Workers, materials, and equipment must be on the site at the right time. Workers are directed to complete the tasks that were scheduled. Construction supervisors and inspectors control the project work. They make sure things are being done right and on time.

KINDS OF CONSTRUCTION PROJECTS

Construction projects vary greatly both in size and kind. The project can be a one story home or a 100 story skyscraper. It can be a new road or a golf course. It can be a new sewer system or a power plant.

There is an easy way to study the many kinds of projects. All construction projects are put into two groups: building construction and heavy engineering construction.

Building Construction

Building construction ranges from a home to a skyscraper. Buildings can be placed into five groups. Included are houses, public service buildings, commercial buildings, high-rise buildings, and factories. Fig. 2-7 shows one of each kind.

Houses
Construction companies of all sizes build houses. A one-person company may serve as a general contractor. The *general contractor* is responsible for the entire project. This general contractor then hires *subcontractors* to do special jobs. Other construction companies are large. They may build entire housing developments, apartment buildings, and condominiums. A condominium is an apartment building where each apartment is owned by the people who live in it.

Fig. 2-7. This group of pictures shows building construction projects. Do you know which kind of building each picture represents? (A–Elliott Corp.; B,C,D–Pella/Rolscreen Co.; E,F–Libbey-Owens-Ford Co.)

Public service buildings

Schools, courthouses, and hospitals are public service buildings. Public buildings are more complex to build. They are built by people who put up only one kind of building. Small contractors do not often put up public buildings. These buildings may require a great deal of large equipment and special installations.

Commercial buildings

Commercial buildings include office buildings, supermarkets, shopping centers, and banks. They are built to house firms that sell to and serve the public. A major goal is low cost space. To help keep costs down, the buildings are often one story buildings. The building is simple. Cranes are seldom used.

Industrial buildings

Industrial buildings include small factories and warehouses. These buildings are unlike retail structures. The projects have many special needs. Their power needs may be huge. The factory process and the concern for the worker are the prime factors when building industrial buildings.

Highrise buildings

Highrise buildings house offices, apartments, or hotel facilities. Highrise buildings are often built in cities where land is scarce and costly. Moving materials over busy streets takes time and is costly. The taller the building the more it costs to build. It takes more equipment, time, and work to get materials and workers to the level where work is done. However, the cost is not high when compared to the initial land cost.

Heavy Engineering Construction

Construction projects which are not buildings are called *heavy engineering construction.* The projects in this group include roadways, water control projects, piping systems, industrial complexes, and electric power and communications systems.

Roadways

Roadway construction has made our transportation system better. Roadway projects include railroads, highways, service roads, and airport runways.

We are able to travel a long way quickly and safely because of more and better roadways. See Fig. 2-8. Roads are built to go around, over, or

Fig. 2-8. Highways stretch in all directions through our nation. (Asphalt Institute, CMI, Arizona Department of Transportation)

through obstacles. A bridge is used to carry a roadway over something such as a body of water or other roadways. See Fig. 2-9.

When it costs less to build a roadway below ground, tunnels are dug. See Fig. 2-10. Subways are railroads that run in tunnels beneath city streets.

Water control projects

Water control projects restrain the force of water in oceans, lakes, and rivers. *Jetties* are walls of rock or concrete used to turn aside currents and tides. A *breakwater* is a barrier that protects the shore from the full force of waves. *Dredging* is used to deepen harbors and rivers. Power shovels and suction dredges remove sand and soil and deposit it elsewhere.

Dams are built to control rivers, Fig. 2-11. Water is held back at flood time to prevent flooding down-

Fig. 2-11. Dams control rivers, provide electric power, and store water. (Associated General Contractors of America)

Fig. 2-9. Bridges carry a road over an obstacle. (HNTB Engineers)

Fig. 2-10. Tunnels shorten the route for roads. (Pennsylvania Department of Transportation)

stream. Water is also held back to supply drinking water and water for crop land, to generate electricity, and to provide play areas. The engineering and building of a dam takes years.

Pipelines

Pipelines bring natural gas, oil products, and water to us. See Fig. 2-12. A sewer is a pipeline for taking wastewater and other sewage away. Pipelines are most often placed underground. The ground protects and insulates the pipe. When a pipeline must go over something, such as a river, it is held up by a bridge or trestle.

Industrial complexes

Steel mills, paper mills, oil refineries, and shipyards are very large and complex. They may cover

Fig. 2-12. Thousands of gallons of water per minute will flow through this concrete pipe. (Price Brothers)

hundreds of acres of land. Planning is as important as construction for these projects. The plan must consider storage areas and efficient transportation of materials.

An electrical power system consists of a generating plant, switchyard, transformers, and distribution lines. See Fig. 2-13. The electricity is produced in the electrical power plant. The power plant, switchyards, and transformers are all part of the structure. The lines and towers carry the power to distant places. Wood, concrete, or metal towers are used to support the power lines.

The building of electrical communications systems is much the same as for electrical power systems. See Fig. 2-14. Telephone, telegraph, radio, television, and navigation equipment are kinds of electrical communications systems. People with special skills install most communication equipment.

PURPOSES FOR CONSTRUCTION

Take a moment and look around. What construction projects do you see? What purpose does each serve? If we made a list of them, it would be a long one. The list of purposes can be made simpler. Most of the items can be placed into six broad groups. The basic purposes are to contain, shelter, manufacture, transport, communicate, and provide recreation.

Contain

A structure may be built to contain something. To contain something you must hold it inside a space. A water tower or dam contains water until it is needed. Natural gas is held under great pressure

Fig. 2-13. Generator plants have been built throughout the country. Power lines carry electricity all over our nation. (Bob Dale)

Fig. 2-14. Microwave towers are used to transmit telephone messages.

in huge ball-shaped tanks. Grain is stored in bins. Fences contain livestock.

Shelter

People and products need shelter from the forces of nature. See Fig. 2-15. These forces include snow, rain, wind, heat, and cold. Shelter holds the forces of nature outside of a space. Living things are more healthy if they have shelter from the cold (or heat) and rain. Barns provide shelter for livestock. A greenhouse protects plants. People build houses so they can be warm and dry.

A water tower keeps water clean. A warehouse protects products of all kinds.

Manufacturing

Many projects are built to manufacture a product. Building glass plants requires a lot of land. See Fig. 2-16.

Fig. 2-16. This site has a glass plant on it. Many structures and a lot of land are used to produce glass. (Libbey-Owens-Ford Co.)

Fig. 2-15. This structure shelters people, production equipment, and finished frames for buildings. (Stran Buildings)

Transport

The purpose for roadways and piping systems is to transport people and products. Highways make it easy to move people and products over a long or short distance.

Piping systems transport solids, liquids, and gases. Most piping systems have gathering pipelines, a processing plant, and distribution pipelines. In the oil industry, the system starts at the well. Other piping systems handle waste. Wastewater enters the system at the home or at the industrial site.

Electrical power is sent along huge metal cables held by towers. The cables are a part of an electrical power transmission system.

Communication

Construction projects built for communication are common. See Fig. 2-17. Information is sent and received by telephone, telegraph, radio, and television. Even road signs are built to provide people with information.

Recreation

A golf course, baseball park, or theater is built for recreation. Fig. 2-18 shows another project for recreation. These projects provide a place for people to play and watch others perform. They provide space to exercise and to sit.

CONSTRUCTION PROCEDURE

Construction projects all involve the same steps. The project is planned, built, and serviced. Fig. 2-19 suggests steps in the construction cycle. Steps from design through maintenance are shown in the figure.

Planning

Planning begins when a person has an idea and gets the decision making started. This step is called *initiating.* Initiators determine the need for the project. They may have to convince others of the need.

A *feasibility study* is done to see if the project is a good idea. A clear recommendation comes out of the feasibility study.

If the owner accepts the project, people start to design and engineer the project. Next, they select the site.

Fig. 2-17. Projects do not have to be large. (Birdview)

Land and soil surveys are made. A *survey* is a process used to measure and describe the land. Soil testing is done to see if the land can support the weight of the structure. When a site is found to be suitable for the project, the land is purchased.

Most of this information is used in the design and engineering stage. In this stage, the working drawings and specifications for the project are made. See Fig. 2-20. *Working drawings* describe the placement, shape, and size of the roof framing. The

Fig. 2-18. This basketball stadium is for the enjoyment of athletes and spectators. (TEMCOR)

THE CONSTRUCTION CYCLE

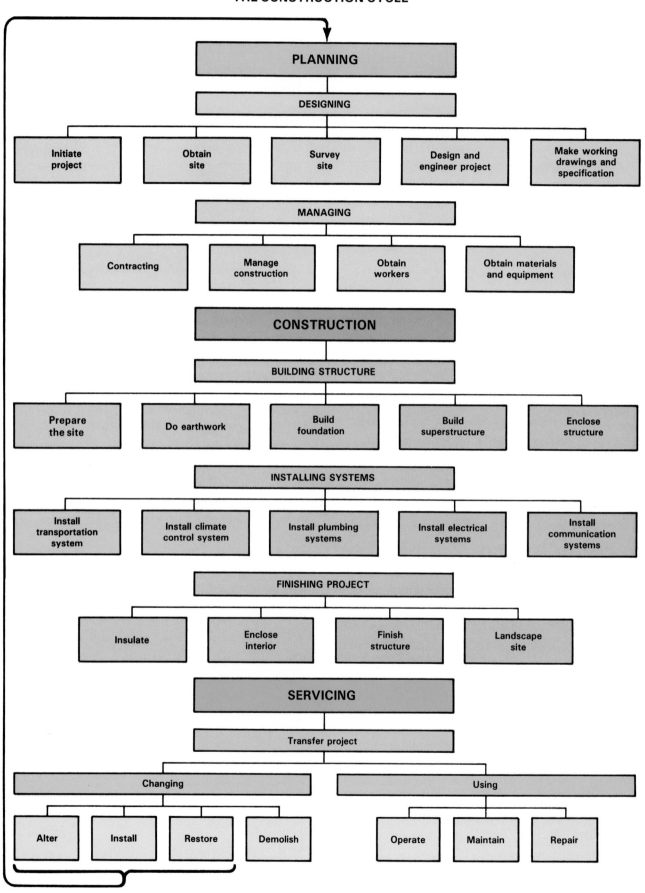

Fig. 2-19. Some steps in the construction cycle.

Fig. 2-20. Much designing is done with computers. (Stran Buildings)

specifications describe the kind of materials and the quality of work the owner expects from the builder.

The success of the project depends on the builder. The builder must be able to do good work. This means doing it on time and at a reasonable cost. Builders, who want the job, use working drawings and specifications to prepare *estimates* (probable cost). The estimates are given to the owner as a *bid.* The contract is often given to the lowest bidder.

Building

The second phase, building, begins when the contractor starts to prepare the site. In this process, the site is cleared of features that will not be used. Fig. 2-21 shows two ways that sites are cleared. Temporary structures and service roads are built.

Earthmoving machines are used to change the shape of the earth's surface. See Fig. 2-22.

When building a structure, you start from the bottom and work up. The *foundation* connects the structure to the earth and supports its weight. The foundation may consist of steel beams, wood poles sunk into the ground, and concrete or wood walls.

Superstructures with steel, concrete, and wood frames are built on top of the foundation. The out-

side is enclosed when shelter is the purpose of the structure.

Next, the mechanical systems are put into the structure. The *mechanical systems* include electrical, plumbing, heating/cooling, and communication. In large buildings, mechanical systems may include conveyance systems.

When you enclose the inside, you add insulation. Then you enclose the walls, ceilings, and floors. Next, the builder paints, decorates, trims, and installs special equipment. The outside involves roofing, siding, and masonry. *Landscaping* the land outside includes grading, planting, and installing outside features like driveways or curbs.

Fig. 2-21. Building begins when the builder starts to clear the site. It ends when ownership is transferred from the builder to the owner.

Fig. 2-22. Heavy machines called scrapers shape the earth. (Caterpillar Inc.)

The final building process is *cleanup.* Windows, walls, and floors are washed. Scrap is picked up and hauled away.

Inspection for proper materials, methods, and quality of work is done throughout construction. After the project passes a final inspection, there is a transfer of ownership from the contractor to the owner.

Servicing Projects

The contractor is responsible for solving problems occurring through the warranty period. This period is negotiable (can vary), but one year is common. The contractor may have to start up a factory process and help in the running of new machines. After these agreements are met, the owner is responsible for servicing the project. To *service* a project means to use and to change it as needed. When a project is used, the owner operates, maintains, and repairs it. Changes are made by altering or restoring and by installing new equipment. Fig. 2-23 shows servicing practices. See if you can identify each.

SUMMARY

Construction technology means using resources efficiently to produce a structure on a site. Construction projects are grouped into buildings and heavy engineering projects. Construction projects serve one or more purposes. The purposes are to contain, shelter, transport, manufacture, communicate, or provide recreation. Planning, building, and servicing are the three phases construction projects go through.

Fig. 2-23. Servicing is the responsibility of the owner. See if you can identify the different servicing functions. (Mike Brian, Mudcat, CMI)

KEY WORDS

All the following words have been used in this chapter. Do you know their meaning?

Assemblies
Bid
Building construction
Cleanup
Construction methods
Estimate
Equipment
Feasibility study
Finished product
Foundation
General contractor
Heavy engineering construction
Initiating
Landscaping
Management
Material resource
Mechanical system
Resources
Service
Specifications
Subcontractor
Superstructure
Survey
Water control project
Worker
Working drawings

TEST YOUR KNOWLEDGE

Write your answers on a separate sheet of paper. Do not write in this book.

1. Construction _____ is using materials, labor, equipment, methods, and management efficiently to produce a structure on a site.
2. Signs for a new highway are classified as:
    a. Natural material.
    b. Manufactured material.
    c. An assembly.
    d. A finished product.
3. Construction projects are put into two groups: _____ _____ and heavy engineering construction.
4. A general contractor hires a _____ to do special jobs.
5. The process to be done in the factory and the concern for the _____ are the prime factors when constructing industrial buildings.
6. Which of the following is not labeled heavy engineering construction?
    a. Roadways.
    b. Warehouses.
    c. Water control projects.
    d. Pipelines.
    e. Steel mills.
7. What six purposes are there for construction?
8. To see if land can support a structure, perform a _____ _____.
9. The _____ describe the kind of materials and the quality of work expected from the builder.
10. Arrange the following construction steps in the proper order. (List them by the letter.)
    a. Foundation.
    b. Landscaping.
    c. Prepare the site.
    d. Earthmoving.
    e. Mechanical systems.
    f. Enclose.
    g. Cleanup.
    h. Servicing.
    i. Finish the inside.

ACTIVITIES

1. Watch the newspapers for a week. List the stories that relate to planning, building, or servicing construction projects. Read one that interests you and report on it.
2. Look in the yellow pages of your phone book. Look up Architects, Engineers, Contractors, and other business people in construction. How many of each are listed?
3. As you ride to school, see if you can find construction projects being built. Are they classified as building construction or as heavy engineering construction?
4. Pick any construction project, new or old. Explain the purpose of the project.

Your study of construction technology will probably include practice in drafting. A knowledge of how to read drawings will also be important. (AMCO Elevators)

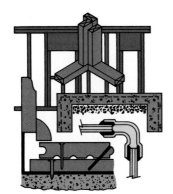

Chapter 3

WHY STUDY CONSTRUCTION TECHNOLOGY?

After studying this chapter, you will be able to:
- ☐ Explain how you can use knowledge of construction technology.
- ☐ Make better decisions about choosing a career.
- ☐ Understand why you should learn about yourself and your aptitudes.
- ☐ Gain more control over future events.
- ☐ Describe how to start a construction business.
- ☐ List requirements for a successful construction business.

This chapter discusses seven reasons for studying construction technology. These are to: learn of its impact, find out how to make better decisions, discover how to be more in control of our lives, learn about work, explore careers, understand yourself by clarifying your goals and aptitudes, and explore entrepreneurship.

IMPACT OF CONSTRUCTION TECHNOLOGY

Construction technology impacts (changes) our lives. Modern structures shelter us from the weather, supply us with power and fresh water, carry away wastes, and allow us to live in any climate we wish. As a further example, temporary housing, Fig. 3-1, impacts both a motel owner and the traveler who needs the shelter.

Construction provides the highways and railroad tracks that let us travel safely to distant places. It provides structures for organizing and directing air transportation, Fig. 3-2. Airplane tires are supported on very smooth surfaces at an airport.

It is construction technology that allows people to enjoy fresh lettuce grown in California or salmon caught in Alaska. Products from all over the world

Fig. 3-1. The need of people for temporary shelters during travel is answered by construction of modern motels and hotels. The people in the picture could be guests, the architects who designed the building, the owners, or the managers. In any case, construction is important in their lives and the lives of travelers. (Jack Pottle)

can be found in your hometown. Transportation systems help to make it all possible.

Modern utilities and services are possible because of construction technology. Beneath the streets of a modern city exists a network of cables, pipes, and tunnels. Fig. 3-3 helps you see what

Fig. 3-2. Modern airports serve thousands of people daily. (HNTB Engineers)

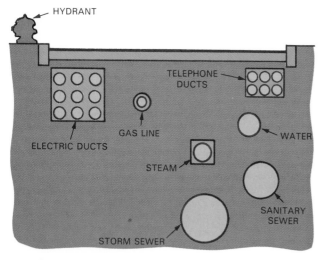
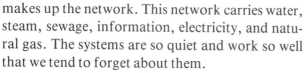

Fig. 3-3. There is a great deal of construction we never see, but we enjoy what it does. Which of these utilities has the most impact on your life?

Fig. 3-4. What is the impact of this dam on the surrounding area and people? (Bob Dale)

makes up the network. This network carries water, steam, sewage, information, electricity, and natural gas. The systems are so quiet and work so well that we tend to forget about them.

Large tunnels carry high speed trains. Many people who live and work in the city ride the trains.

Construction technology provides us with dams, Fig. 3-4. Dams in turn control floods and store water for producing electricity. They provide lakes for fishing and boating. Drinking water and water

for manufacturing is taken from these lakes. All of these uses have changed our lives. They make an impact on us.

HELP IN MAKING DECISIONS

Construction gives us a better life, and more security. But, these things only come to us for a price. Construction may deprive us of good farmland. Other projects may require large amounts of energy to construct or run. A dam may store needed water but may also cover a scenic valley or fertile cropland.

There is often more than one solution to a problem or need. The more you know about construction technology, the better you can make decisions about them. You will be able to think about the good and the bad sides of a proposed solution. A good choice will increase the good effects and reduce the bad effects. Through a study of construction, you may be able to better understand the choices of others.

HOW TO BE IN CONTROL

Do you want to be in control or be controlled? Most of us prefer the first. If you want to be in control, learn as much as you can. The more you know the more you are in control.

Here is a case showing how a person can be in control. Does a person who can ride and repair a bicycle have more control than one who can only ride it? The first can ride it to deliver newspapers, go swimming, or have fun on it. When a tire needs to be fixed, the first person can repair it. The second person can only look at the bicycle. This person must rely on others and is not in control.

Knowing about construction technology is somewhat like knowing how to fix bicycles. Fig. 3-5 illustrates another example. If you know about construction, you will know what to look for when buying a house. You will be able to maintain the home after you buy it. Repairs or changes will be easy to plan. Perhaps you will be able to do some of the work (maybe all of it). By learning more, you are able to rely on yourself more.

Fig. 3-5. The family living in this home saved money by remodeling it. They could do the work themselves because they understood construction technology. (Ceder Shake and Shingle Bureau)

When you can rely on yourself, you gain courage. With courage you can go ahead knowing you can finish the job. Both knowledge and courage together can help you to build a doghouse, clubhouse, or storage shed you have needed.

LEARNING ABOUT WORK

Work means putting out effort to make things better and for a reward. When you mow the lawn it takes energy and thinking in order to do the job right. Mowing the lawn makes your home look better. Your reward for mowing the lawn can come in many ways. You may get paid for the job with money. A good feeling about doing a good job is a reward, too. The person shown in Fig. 3-6 probably works for more than just the money.

Fig. 3-6. What are the benefits from this work? How might the worker be rewarded? (Chevron Corp.)

A lot of effort and thinking goes into construction projects. What work went into building the bridge shown in Fig. 3-7? Even the smallest projects take effort. If you put up a new mailbox, you can get tired because you have to dig, hammer, and think.

Think how much thought and effort went into the interstate highway system. However, the rewards were great. The reward was roads that handle more cars safer and are less tiring to use than the earlier roads. Many people get rewards for building the highways. The workers and managers are paid for their work. Many feel proud of their part in building new highways. The ones who get the greatest reward are those who use the highways. Some people are willing to work for less money so they can have a job that they feel good doing.

Fig. 3-7. This bridge makes it easy to cross a highway in Utah. How many workers did it take? (Utah Department of Transportation)

EXPLORING CAREERS

By studying construction technology you will learn a lot about yourself. On different days you will be an architect/engineer, a construction manager, or a worker. You will be trying many career roles. At one time you will be helping test soil. The next day you may be locating the structure on the site or building a wall.

Clusters

There are many kinds of careers. Careers can be grouped into clusters. *Manufacturing, construction, communication,* and *transportation* are the four *industrial clusters.* The careers described in this book are grouped in the construction cluster. Some examples of work or processes related to several of the four clusters are shown in Figs. 3-8 through 3-10.

Job-levels

Occupations are at different job-levels. The *job-level* is based on the kind of decisions the person makes. The higher that level the more decisions the person makes. The owner can decide on more things than the supervisor. A supervisor decides on more than a worker.

There is another rule for job-levels that is often true. A person with more schooling or who has

Fig. 3-8. Building a communications satellite. These people are in the communication career cluster. (Hughes Communication, Inc.)

Fig. 3-9. Building and repairing a railroad track. This work is included in the energy/transportation cluster. (Association of American Railroads)

Fig. 3-10. Producing vehicle windshields. This work is part of the manufacturing cluster. (Libbey-Owens-Ford Co.)

worked longer will often have the higher job-level. Look at Fig. 3-11. Who makes the most decisions?

Duties

All careers have duties. *Duties* are the things people do while working. Managers plan, schedule, direct, and control the project. There are two kinds of managers. There are the managers in the home office and the managers of the construction project. Managers of the construction project work out of the field office.

Workers are put into craft groups. Steelworkers, plumbers, carpenters, and electricians are four of the groups. Workers on the job site work with people, data (facts and numbers), and things (tools and materials).

Managers work mainly with people and data. Construction workers on the site work with things and with information that relate to construction methods. Fig. 3-12 shows people as they perform duties.

Working Conditions

Working conditions for construction workers can vary. See Fig. 3-13. *Working conditions* include everything that affects a worker on the job. Most construction workers work outside. Tunnel builders and managers work inside. People who install mechanical systems and finish the inside of the structure spend much of their time inside, too.

Many construction jobs are hazardous (risky). Steelworkers may work in high places. Building tunnels under water is risky, too. At times water cannot be kept out of the tunnel. The work area is filled with compressed air to keep the water out. The workers in these areas are called sandhogs.

Fig. 3-11. Which person probably has the highest job-level? Why? (Honeywell)

Fig. 3-13. Describe the differing working conditions for these workers. (Bob Dale, Utah Department of Transportation)

Fig. 3-12. These workers are performing general duties. Are they working with people, data, or things? (Spectra-Physics, Honeywell)

They can work for only short periods of time. Some people prefer the risky jobs.

Some jobs are very structured. This means their tasks are done over and over again. In most cases the job must be done in only one way and at one place.

Construction workers do more than one thing. Even laying bricks varies. Doors, windows, and corners involve careful measuring work. The measuring work gives variety to the job. Each job and each wall is a new challenge.

Construction workers work in many places. Dams, roads, and electrical power plants may be built in remote areas. Construction workers travel to different towns, states, and countries. Some people like to work in other places. They feel that it is interesting.

Personal Requirements

Personal requirements are traits you have. Kinds of traits include: physical (refers to strength and speed), intellectual (thinking), and emotional (involving feelings). Some careers require one trait far more than another. Look at five examples related to construction as an occupation. Steelworkers need a good sense of balance. Concrete workers need strength. Engineers need math skills. Architects need to be creative. Physical strength and skills with measuring tools and other tools are needed in construction. Knowledge of personal requirements for a job can help you decide which job you can succeed in.

Pre-entry Preparation

Pre-entry preparation will often be needed to get a job in construction. A trade school or college may provide *pre-entry training.* Workers may go to vocational schools. Managers often go to a college. Doing work that relates to a job is called getting *experience.* Working for parents or neighbors is one way to get experience, Fig. 3-14.

A process of testing a worker's knowledge and skills, called *certification,* may be needed. Architects and engineers must pass a test and get work experience to become certified. Skilled tradespeople have to earn a journeyman's card. Each tradesperson must serve as an apprentice, take special classes, and pass a test to get a card.

Rewards

Rewards from work may be in the form of earnings, security, and satisfaction. See Fig. 3-15. *Earnings* come in the form of cash and fringe benefits.

REWARDS AT WORK		
EARNINGS FROM...	SECURITY IN...	SATISFACTION FROM...
• Wages • Benefits • Overtime	• Future Trends • Advancement • Good Retirement Plan	• Taking Charge • Recognition • Achievement • Work • People • Building Projects

Fig. 3-15. There are many kinds of rewards.

Fig. 3-14. Any kind of work in construction gives you experience. What kind of experience are these people getting? (The Davey Tree Expert Co.)

Fringe benefits include health care, retirement plans, and days off with pay.

When workers have little worry about keeping their jobs, they have *security*. Some people will not take a new job for more money if they have a stable job. Do you think it is a good or bad idea to pass up chances to earn more money if you have security now?

Satisfaction can come from building something that the person is proud to build. Some people like to be in charge. Others like attention. How do you get satisfaction?

UNDERSTANDING YOURSELF

The better you know yourself, the better career choices you will make. Learn more about yourself and control your life better. A good question is: "What is there to know about one's self?" The answer has two parts.

Goals

First, people have goals, Fig. 3-16. *Goals* are things you want to achieve. A person may want to achieve a given job-level. Others may want to perform certain duties. Special working conditions may be a person's goal.

A level of income or security can be a goal. Having time and money to pursue a hobby or to spend with the family may be another's goal.

People have many goals. The better you know your goals, the easier they are to achieve.

Aptitudes

Second, *aptitudes* are talents you are born with or develop. People are able to develop many talents they need or want. You may be naturally strong, or you can develop strength. The same is true in learning things. You can probably learn something about construction and mathematics if you want to learn. The people shown in Fig. 3-17 are improving their designing skills so they can become architects.

Just as you can learn skills with tools and data, you can learn skills in leading people. It would be nice if we were born with good aptitudes. It does not always work that way. Some of us need to develop them ourselves. When we know our strong points we can apply them. If we have a weak point, we can build it up. Learning to build up weak points helps us adjust. People who can adjust have more chances to succeed.

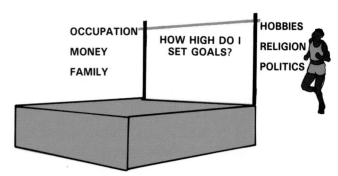

Fig. 3-16. What are your goals? The sooner you decide on your goals the sooner you can begin reaching them.

Fig. 3-17. These people are working to develop their aptitudes. It requires some effort. If it makes it possible to do what you want to do, the effort is worth it. (Honeywell)

EXPLORING ENTREPRENEURSHIP

An *entrepreneur* is a person who starts and manages a business. The entrepreneur assumes a risk to make a profit. Look at Fig. 3-18. There are three roles that an entrepreneur may play in construction. They are project owner, designer, and contractor. See Fig. 3-19. Entrepreneurs can own and manage the project after it is built, provide an architectural/engineering service, or build projects.

A builder may build an office building and then rent and service the space, Fig. 3-20. Some office workers may start a consulting firm and provide engineering services. They may do soil surveys or inspect bridges. See Fig. 3-21.

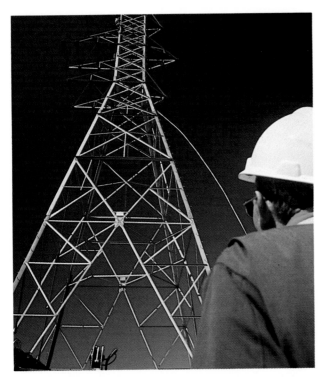

Fig. 3-18. This entrepreneur is risking his money to build a power transmission line. If he is a good leader he will make a good profit. (Bob Dale)

Contractors

Entrepreneurs who build projects are called contractors, Fig. 3-22. They manage shaping the land and building the project. Many areas of work require special training and licenses.

Starting and running a business requires many skills and resources, Fig. 3-23. To start a business in home building you need:

Fig. 3-20. The owner of this builder hopes to rent space for offices. (HNTB Engineers)

- Working capital (money, tools, and equipment).
- Knowledge of construction.
- Some skills in running a business.
- A desire to work hard and long.

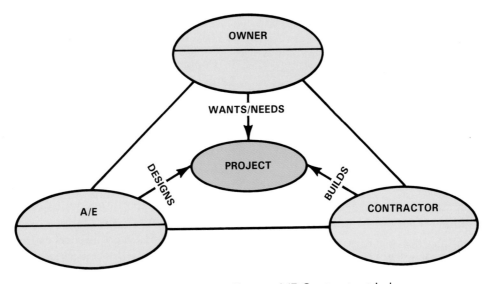

Fig. 3-19. The Project Owner-A/E-Contractor triad.

Fig. 3-21. Special services are provided by entrepreneurs such as this soil surveyor. (Western Technologies, Inc.)

Risks in Construction Business

The construction business is risky. The amount of work available changes more than in other businesses. At one time there may be a lot of work. At other times, there may be very little work. The reason is involved with a term called investment.

Fig. 3-22. The welder is an entrepreneur. He owns a truck, welding equipment, and tools. The helper often is a partner. They may hire on as a team for a general contractor. (Shell Oil Company)

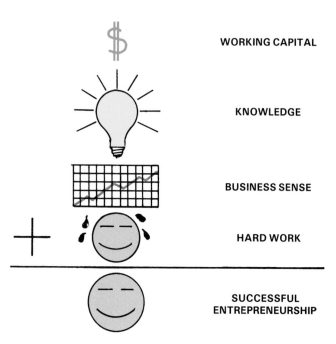

WORKING CAPITAL

KNOWLEDGE

BUSINESS SENSE

HARD WORK

SUCCESSFUL ENTREPRENEURSHIP

Fig. 3-23. The formula for success in construction is the same as in other businesses: resources and hard work.

Money used for building is an investment. An investment is used in the future. A new factory building is an investment built for future production. A home is also built for future use. When a company is not selling all they can produce they do not build (invest in) a new factory. It is like that in a family, too. If your father or mother loses his or her job, would they build a new home? Chances are they would not. The money for the new home would be saved for expenses. Expenses are day-to-day things you need. Unlike expenses, investments are often seen as luxuries. For this reason, investment money is the first to be reduced in times of trouble.

Types of Managers and Workers

People in construction can own their own firms. In construction there is less need for equipment, buildings, office space, and full-time workers than in manufacturing. Though it is easy to start a business, there is a big risk. The failure rate is high.

The move from a worker in the office or in the field to an owner is made in many ways. A person may have been a supervisor (once called a foreman), or job superintendent (leader at the job site) who went into contracting. Some contractors started as clerks who learned estimating and purchasing. Others started with college training in architecture, engineering, construction, or business administration.

General Contractors and Small Builders

Construction is done by these three groups:
- The general contractor.
- The custom builder.
- The upkeep and improvement builder.

General contractors are placed into two groups. They are building contractors and heavy engineering construction contractors.

The easiest field to enter is building. It requires the least money. A *building contractor* often starts building homes or small retail buildings. This person is often a small entrepreneur. The average builder puts up less than 20 units (buildings) per year.

Building contractors may be speculative builders. They build houses hoping to sell them at a profit.

The second group of general contractors, the *heavy engineering construction contractors,* often get work from governments and institutions. This work can be more steady than that for the building contractor.

A *custom builder* constructs what the owner wants built. A future owner calls the custom builder. The owner and builder discuss the project. Then the owner asks the builder to quote a price for building the project. If the price is right, that builder gets the job.

The *upkeep and improvement builder* services existing structures. Contractors may be called to do maintenance (such as painting) on a building. Repair businesses are often small. They may do one kind of work (such as roofing) or handle more than one kind of repairs (such as painting, roofing, and siding). Updating and remodeling jobs are a good way to get started because the jobs are often small. Estimating is hard. Therefore, the risk is higher than when building a new project, Fig. 3-24.

SUMMARY

The following are some reasons for studying construction technology:
- To learn of its impact on us.
- To help in making decisions.
- To put us more in control.
- To learn about work.
- To explore careers.
- To help gain knowledge about our goals and aptitudes.
- To explore entrepreneurship.

Construction involves the use of tools, materials, capital, skills, and management methods to build structures for transportation, manufacturing, and communication. You learn about some of the previous seven areas with familiar examples of your

Fig. 3-24. Remodeling is a way to start a building business. The jobs are small but estimating is hard. (Sellick Equipment, Limited)

successes and failures from past experience and new choices seen from structured study.

Learning about construction technology has value to you. You learn how construction makes your life better and how to make better decisions. You gain knowledge in order to rely on yourself more. From laboratory activities, you learn about the rewards of work and you try out careers. Some of these careers may be what you would like to do when you are out of school. Perhaps you will learn enough to begin planning for a future business which you want to pursue.

KEY WORDS

All of the following words have been used in this chapter. Do you know their meaning?

Aptitudes
Building contractor
Certification
Custom builder
Duties
Earnings
Entrepreneur
Experience
Goals
Heavy engineering construction contractor
Industrial cluster
Job-level
Personal requirements
Pre-entry preparation
Pre-entry training
Rewards
Satisfaction
Security
Upkeep and improvement builder
Working conditions

TEST YOUR KNOWLEDGE

Write your answers on a separate sheet of paper. Do not write in this book.
1. List some types of tunnels under a city street.
2. The more you know about _____ the easier it is to buy a house.
3. Careers in construction include architect/engineer, _____ _____, or worker.
4. What are the four career clusters?
5. Which workers do their job inside?
   a. Steelworkers.
   b. Tunnel builders and managers.
   c. Carpenters.
   d. Electricians.
   e. Surveyors.

Matching questions: On a separate sheet of paper, match the following classes of workers with the traits they need for skillful and safe work.
6. _____ Concrete workers. a. Math skills.
7. _____ Architects. b. Strength.
8. _____ Engineers. c. Balance.
9. _____ Steelworkers. d. Creativity.
10. Some _____ preparation at a trade school or college is needed to be certified for a job in construction.
11. Self-knowledge means knowing your _____ and your aptitudes.
12. An entrepreneur assumes a _____ to make a profit.
13. True or false? It requires less money to enter the field of heavy engineering construction than it does to enter building construction.

ACTIVITIES

1. Write down what life would be like if we did not have things made by construction.
2. If you were allowed to rebuild your school or build it as a totally new structure, how would you change the way it was built?
3. What areas of your life would you like to control? What would you have to learn to gain control?
4. Write down what you think about work. What is work? Why do we work? Should we have to work?
5. Visit a neighbor. Ask your neighbor where he or she works. Ask questions about the person's job-level and other career characteristics. Report to the class what you found out.
6. Describe the ideal job for yourself.
7. What are your goals? What would you like to achieve?
8. Think about being 25 years old. Describe a weekday from the time you get up until you go to bed.

Knowledge of what tools are available can save much work. This is true for construction jobs and other jobs. (Black & Decker Industrial/Construction Division)

Section 2
PREPARING FOR THE CONSTRUCTION PROJECT

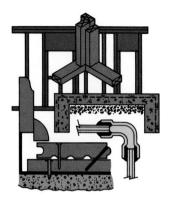

Chapter 4

INITIATING THE PROJECT

After studying this chapter, you will be able to:
☐ Describe what an initiator is.
☐ State the purpose of a public hearing.
☐ Define the term "cost/benefit analysis."
☐ Explain the need for consultants and a professional staff.

Before anything is constructed or even designed, someone must need or want the project. Projects built to improve the quality of life are said to be needed. Safer highways, better wastewater treatment plants, and nicer parks are three that improve quality of life. A new highway saves lives and time for travel, Fig. 4-1. The quality of water in rivers is improved by modern wastewater treatment plants, Fig. 4-2. Parks provide attractive areas where people can relax and enjoy themselves.

WHO INITIATES?

Monuments and expensive buildings are often built to satisfy people's wants. Monuments draw people's attention to a person or idea. Expensive buildings provide attention, prestige, and importance to a person, firm, or city.

The person or group that has the idea is the *initiator.* This person or group must get things started. See Fig. 4-3.

Fig. 4-1. This highway makes travel to the heart of the city easy. (Asphalt Institute)

Fig. 4-2. This wastewater treatment plant makes the water safe to put back into the river.
(DYK Incorporated)

INITIATING PRIVATE PROJECTS

Initiating a project follows a procedure. It is shown in Fig. 4-4. The process starts with an idea and ends with a decision. The decision is to build or not to build.

Initiators

In **private projects,** the initiator and the owner are usually the same person. Look again at Fig. 4-3. Private initiators want something to benefit themselves. It may be a factory to produce a product and

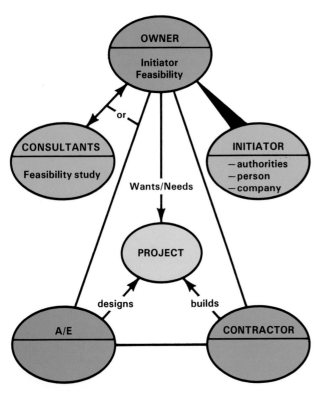

Fig. 4-3. The owner is the focus when a project is initiated.

Fig. 4-4. There is a process for initiating a private project.

profits. A new office building may improve the company's image. A new roadside sign may increase the number of guests staying at a motel. A person may build a house in which to live. Another may build it to sell for a profit. All of these projects benefit the initiator. They provide the initiator profit, image, or convenience.

Obtain Funding

The private project initiator must pay for it. The money must be in hand or be borrowed. It is the initiator's job to convince the people who lend money to invest in the project. If the initiator works for a company, the leaders of the company need to be convinced the project has value.

Select Professional Staff

The next step is to select the *professional staff,* the people who will do the thinking work behind the project. Large firms may have their own. The owner of a small firm may conduct the feasibility study. Most firms hire consultants. *Consultants* do research and make recommendations on how to design, finance, manage, and promote the project.

Study the Problem

The consultants study all reasonable ways to solve the problems. To solve the problems they need to collect information. For example, a restaurant owner, who wants to build on another site, needs to know how the future site is zoned (residen-

tial, commercial, industrial, or agricultural). Vital facts include:
- Where are the potential sites?
- How much pedestrian traffic is there?
- What kind of traffic is there?
- How much auto traffic is there?
- Is it easy to get to and from the restaurant?
- What is the economic condition of the surrounding area?
- Is the neighborhood growing?
- What are the neighborhood residents' habits for eating out?
- Where do they eat out?
- How many other restaurants are there?
- What kind of restaurants are they?
- How well are the restaurants doing?
- How much will it cost to build?
- How much parking is needed?
- What person or institution will supply the money?
- Are workers available?

Technical facts are found out about the site. These facts include:
- The kind of zoning.
- The kind of soil.
- The contour of the land.
- The kind of utilities.
- The location of the utilities.
- The property lines.

The owner and the consultants then study the facts. They compare the cost of the project at the different sites. This comparison is a *cost/benefit analysis.* The owner is looking for the most benefit at the least cost. Fig. 4-5 shows a cost/benefit analysis for a restaurant.

	SITE A		SITE B	
Land Cost	$100,000	5,000/yr.	$ 80,000	4,000/yr.
Building Cost	80,000	4,000/yr.	100,000	5,000/yr.
Taxes		2,000/yr.		1,700/yr.
Labor		50,000/yr.		45,000/yr.
Food Costs		120,000/yr.		108,000/yr.
Building & Op. Costs/yr.		181,000/yr.		163,000/yr.
Potential Sales		197,300/yr.		180,000/yr.

$$\frac{\text{Potential Sales/yr.}}{\text{Building \& Operating Costs/yr.}} = \frac{\text{Return for}}{\text{Investment}}$$

$$\frac{\$197,300/\text{yr.}}{\$181,000/\text{yr.}} = 1.09 \qquad \frac{\$180,000/\text{yr.}}{\$163,000/\text{yr.}} = 1.104$$

Fig. 4-5. A cost/benefit analysis for a restaurant. Which site seems to be the best to you?

Make a Recommendation

Finally, the professionals make recommendations, Fig. 4-6. The *recommendations* state what, where, how, and when the project should be built. In the restaurant example, the recommendation would include:

- The specific location.
- The kind of building.
- The menu and price.
- How the food should be served (drive up, fast food, sit down).
- The amount of parking.
- Should it be franchised or not.

Make a Decision

The next step is to make the decision to build or not to build. If the decision is to build, they must further decide what is to be built.

These decisions to build and how to build are made by the highest level managers.

Management may or may not follow the consultants' recommendations. For example, the consultants recommended not to buy a franchise. The owners may decide to buy the franchise anyway. They will need more money, but they feel the benefits will be worth the cost. They will get many things in return for the extra money. The franchise will provide national advertising, assistance in designing the building, a set menu, and a source for food. The workers will receive training in how to prepare and serve food. The manager will receive training in how to supervise people and manage a business.

INITIATING PUBLIC PROJECTS

Public projects are funded by local, state, and federal taxes. The purpose of public projects is to benefit the *general public* rather than a person or firm. However, a person or firm often gains. Those who plan, build, and service the project will be paid for it. Schools, highways, sewers, and water treatment plants are public projects.

Initiators

Public projects are initiated to help the economic growth of the city or area. Fig. 4-7 shows part of a public transportation system. Fig. 4-8 shows an airport for public use. Business tends to go to cities that have good *infrastructures* (schools, streets, transportation, utility systems, and parks). Mayors and city councils often initiate public projects that help economic growth.

Public projects are at times mandated (forced) by regulatory authorities. A *regulatory authority* sets standards that a city or company is to follow. The state board of health is one. If the quality of water is poor, a new water treatment plant may be mandated.

A group of citizens may initiate a project. The group may file a civil action (or threaten to) against the city because a stream smells bad. This action

Fig. 4-6. The owners and the consultants meet to discuss the recommendations. (Stran Buildings)

Fig. 4-7. Public transportation systems make it easy and cheap to travel in cities. (HNTB Engineers)

Fig. 4-8. A medium size airport. When the airport was initiated, many people attended the public hearing. (Hellmuth, Obata and Kassabaum, Architects)

may result in the initiation of a wastewater treatment plant project.

Obtaining Support

The initiator tries to get support. Publicity is used to inform people of the benefits of the project. A public hearing may be called. At a ***public hearing*** the initiator states the reasons why the project should be built. It takes a lot of work to gain support for public projects.

Getting a Professional Group

The next step is to select the professional group. The ***professional group*** is made up of architects, engineers, financial experts, and estimators. They design and see that the project is built right, and that the project works. Training is included when special equipment is installed. Those who wish to be considered present their credentials (qualifications) to the city leaders. The engineers should be familiar with the technical problems, and with the area. Other experts know the ways to get funds and the feelings of the people.

Studying the Problem

The professionals conduct the feasibility study. They consider alternatives. They gather all impor-

tant facts. The facts are reviewed. Finally, they make a recommendation to responsible members of the city government.

Who attends the planning meeting is determined by the project. An executive from the city government (usually the mayor) and the city council attend, Fig. 4-9. If the leaders are not able to attend, representatives of the group attend. If a park is being built, the director of parks attends. The sanitary district members attend meetings that review wastewater projects.

Fig. 4-9. Large projects require that many people attend planning meetings. (HNTB Engineers)

Making the Decision

The city leaders then decide on the future of the project. They decide to continue (go on with) or stop the project. If they will proceed, they decide the future steps for the project.

SUMMARY

The initiations of private and public projects are quite similar. Someone gets an idea and informs others about it. Professionals collect data and ideas. Recommendations are made. Top level leaders decide on whether to build or not to build. They also decide the general idea for the project.

Initiating private and public projects differs in some ways. Private projects serve self-interests. Public projects serve community interests. Initiating private projects involves few required procedures. Initiating public projects follows a set process. The process is stated in local, state, and national laws.

People study and decide upon public projects at public meetings. Anyone can attend.

KEY WORDS

All the following words have been used in this chapter. Do you know their meaning?

Consultant
Cost/benefit analysis
Infrastructure
Initiator
Private project
Professional group
Professional staff
Public hearings
Public projects
Recommendations
Regulatory authority

TEST YOUR KNOWLEDGE

Write your answers on a separate sheet of paper. Do not write in this book.

1. The person or group that has the idea to build is the _____.

2. Which of the following can act as a public initiator?
   a. A profit-making company.
   b. A city council.
   c. A homeowner.
   d. A social club.
   e. A concerned group of citizens.

3. A _____ study shows whether a project can be built for a given cost.

4. Most firms hire _____ to conduct a feasibility study.

5. An area may be _____ residential, commercial, industrial, or agricultural.

6. List some questions the restaurant owner should ask about a neighborhood.

7. A _____ states what, where, how, and when the private project should be built.

8. Arrange in order the following steps for starting a public project:
   a. Study the problem.
   b. Obtain support.
   c. Get a professional group.
   d. Initiator's action.
   e. Make a decision.

9. A _____ _____ is a meeting to gain support for a public project.

10. A _____ group is made up of architects, engineers, finance experts, and estimators.

ACTIVITIES

1. Choose a new private construction project. Call the owner on the telephone. Ask the owner, "Why did you build the project? Why was it built were it was?" Report your findings to the class.

2. Select a recent public construction project. Call the administrative unit on the telephone. Ask a person in authority, "Why was the project built? Why was the project built where it was built?" Report your findings to the class.

3. Identify a problem in your city or town. Discuss ways to solve the problem.

4. Go to the community planning office. Ask someone there, "How do you decide which railroad crossings to repair first? How do you decide where to place street lights?"

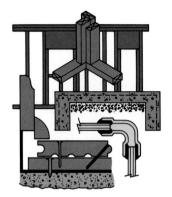

Chapter 5
OBTAINING THE SITE

After studying this chapter, you will be able to:
- ☐ Discuss site selection by a mayor or company president.
- ☐ Tell how costs and building codes affect the site selection.
- ☐ Define site purchase by negotiation or through condemnation.
- ☐ Explain the purpose of a mortgage.
- ☐ List ways to create legal descriptions.

After the decision to build is made, a site must be obtained (bought). The owners or city leaders need to select the site that best serves the purpose, Fig. 5-1. Studies of site costs, construction costs, and usefulness of the site help in the selection.

In the next step, the initiator buys the site. A building permit is usually obtained after buying the site. In this chapter, selecting and buying the site are described in more detail.

SELECTING THE SITE

The key to success for any construction project is to select the best site. The goal is to get the lowest overall cost for the highest value. Fig. 5-2 shows a way to figure costs.

A factory needs to be close to its raw materials, labor, and markets. Schools should be near the middle of the district. Monuments should be where people can see them.

Fig. 5-1. Why do you suppose the owner chose this site? (Cedar Shake and Shingle Bureau)

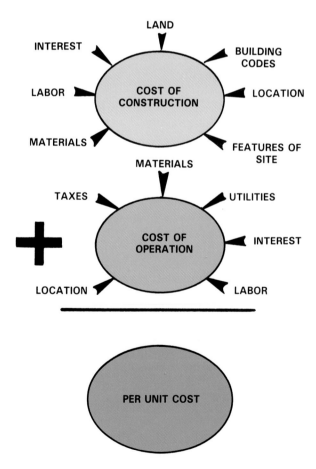

Fig. 5-2. Per unit cost formula.

Fig. 5-3. A bridge is not always placed at the most narrow point in a river. This bridge needs rock ledges for support. It is, also, in the direct route for the road.
(U.S. Department of Transportation, Federal Highway Administration)

Much care and time is used to select the site, Fig. 5-3. After building the dam, highway, or pipeline, it is hard and costly to move it. Manufacturers seldom move their plants. Schools are used until they are too costly to repair. Then new ones are built.

Who Selects the Site?

Sites for private construction projects are selected by top management. The owner decides on the proper site.

Public projects require a different process. The mayor, city council, and responsible administration (such as a school board) decide on sites for public works.

How are Sites Selected?

Recommendations were a part of the feasibility study. Based on these recommendations, professionals gather facts from more than one site. A report that sums up the facts on each site is made to those who decide. Those who decide visit the best sites among the many sites listed.

After the visit, the number of sites is reduced to a few. A careful study is made of each site. The owner or the government leaders select the one best site.

The impact of the project on the environment is a critical concern. An ***environmental impact report*** may state that construction will negatively affect

endangered species of plants or animals. Some sites are rejected for this reason, Fig. 5-4.

The cost of the land is an important factor. This is most important when much land is needed. Some highway projects use thousands of acres of land.

The goal is to get the greatest value for the purpose of the project, even if land and building costs are higher. Owners and civic leaders try to select

Fig. 5-4. Some structures can only be built in one place. What effect would an oil spill have on the environment? (Combustion Engineering, Inc.)

sites that will reduce the cost of construction and the cost of operation. If they succeed, the site will reduce the cost per unit to produce the product or provide the service.

Cost of Construction

Several factors affect the cost of construction. The first factor is the price of the land. There are two big questions:

- Will the owner sell the land?
- What is the price?

When a large area of land is needed, the second question is more important than when a small amount is needed. The price for land is of more concern to some owners than others. A person building a country club is more concerned than a person building a miniature golf course.

There are laws that control how people use the land. *Building codes* are standards (rules) that protect people. Building codes keep structures in agreement with zoning ordinances (laws). People enact zoning ordinances to control land use. *Zoning ordinances* describe how structures are used, occupied, and placed on the land.

If the land is not zoned for the planned use, zoning must be changed. Most building codes describe four major zones. They are:

1. *Residential*–Where we live.
2. *Commercial*–Where we buy, sell, and get various services.
3. *Industrial*–Where we produce products.
4. *Agricultural*–Where we grow food and other materials.

The standards differ for each of the zones. For example, one residential zone allows only single family dwellings. Another zone allows duplexes (two family dwellings). Multiple family dwellings (apartment houses) can be built in other areas zoned for it. The zoning code protects people. A house in a single family zone cannot be converted into an apartment house unless the zoning is changed. Fig. 5-5 is a zoning map for a section of a city. Changing the zoning takes time and costs money. The harder it is to change, the more the land costs.

Codes control the way buildings are to be built. Codes set standards for the design and materials used for the project. Different areas have different building codes. Where earthquakes are common, the height of a building is limited. Other areas limit the amount of land that the building can cover.

Therefore, building on one site may cost more or less than building on another site.

The list of features on the site affects the cost of construction. What is on the surface? Are there buildings, trees, and rocks that need to be removed? The more there is to clear from the site, the higher the cost of construction.

The shape of the land needs to be considered. How much does it need to be changed? The building cost will be higher on a site where a lot of earth needs to be moved.

A study such as a subsoil survey will show if the soil will support the weight of the structure. The subsoil structure will suggest the kind of foundation to use. A foundation includes the footing and wall above it. This supports the structure. Some foundations cost more to build than others. Building costs go up when the foundation must be set deeper.

The fifth factor in cost of construction is the location of the site. Building highways and power transmission lines is more expensive in isolated areas than others. Materials and workers must be transported to the work site. Electrical power plants take many people many years to complete. If they are built on a remote site, temporary towns may be constructed. Building in cities is hard, too. It is hard to get equipment and materials to a downtown building site. All of the additional work and time raises the building costs.

The sixth factor that affects the cost of construction is the cost of labor. Are workers available? Are they properly trained? What is the pay scale? If labor is not available, others will need to be brought to the site. Higher wages may need to be paid to get workers to come to the job site.

Workers without needed skills must be trained. At first they will be less productive. A less productive person may complete work, but may make mistakes, and may not be able to do certain jobs.

The cost of each hour of work goes up when the cost of getting, training, and paying labor goes up. If the cost of labor rises and if productivity falls, the cost of construction will increase more.

Finally, the cost of getting materials affects construction costs. Are good materials available? Even dirt, sand, or gravel is costly if it has to be hauled a long distance, Fig. 5-6.

Other factors such as the *interest,* money that is paid to borrow a large quantity of money, and the cost to reduce the impact of the project on nature affect the cost of construction.

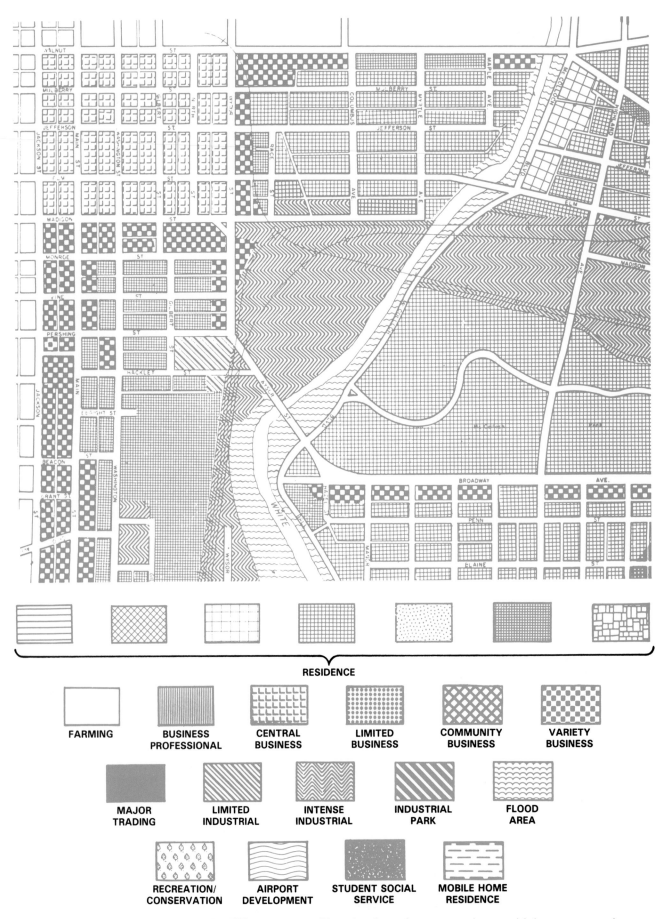

RESIDENCE

FARMING

BUSINESS PROFESSIONAL

CENTRAL BUSINESS

LIMITED BUSINESS

COMMUNITY BUSINESS

VARIETY BUSINESS

MAJOR TRADING

LIMITED INDUSTRIAL

INTENSE INDUSTRIAL

INDUSTRIAL PARK

FLOOD AREA

RECREATION/ CONSERVATION

AIRPORT DEVELOPMENT

STUDENT SOCIAL SERVICE

MOBILE HOME RESIDENCE

Fig. 5-5. Land in a city is zoned in different ways. Top. Look at the map and see which zones are shown. Bottom. The meaning of symbols. (Delaware Muncie Metropolitan Plan Commission)

Fig. 5-6. Special equipment is used to haul large diameter prestressed concrete cylinder pipe. (Price Brothers)

Cost of Operation

The choice of a site often depends on the cost of operation. After the project is built, it must be operated. The cost of operation is greatly determined by the cost for utilities. Gas, electricity, waste treatment, and telephones are **utilities.** Other costs include labor, materials, distribution, and taxes.

First, a site must have utilities. The cost of utilities affects the cost of the product or service. If large amounts are needed, a site should be close to the source. Some firms may decide to provide their own utilities. They may drill wells, keep natural gas in tanks, and treat their own wastewater.

A site that has local labor to run production is better off than one that does not. A fast food restaurant near a housing area will find it easier to get workers than one on the edge of town where there are businesses. Workers on the edge of town will need to drive to work. Higher wages will be needed to pay the worker's travel expense. A larger parking lot may have to be built to provide parking for the workers who drive.

Businesses can lower their operating costs by being close to their supply of materials. Lumber mills are usually placed close to forests. Gypsum board mills are built close to gypsum mines. Even an auto garage can save time by being close to an auto parts store.

It costs a company less to distribute its product if it is close to the market. Short routes result in fast delivery, less packing, and lower transportation costs.

A repair service center building should be placed near the center of the service area. In this way it is easier to get to the service center. When one offers a home service, the service people will spend less time driving, and have more time to provide the service, if they are near the market. By being close to the market, a firm can reduce repairs, accidents, and the number of vehicles for service people.

Local, state, and federal taxes on property and income may affect operating costs. Low taxes reduce operating costs. High taxes raise them. Whichever effect taxes have, they are of concern to the people who select a site. Fig. 5-7 shows an oil port. It is close to the materials needed. Transportation is inexpensive, labor is near, and taxes are low.

After the decisions about the site are made, a **building permit** must be obtained. The city or village building inspector grants this permit after approving the **plan** (drawing) for the project.

Fig. 5-7. Being close to oil reserves, pipelines, and a protected water cove makes this a good site. (Exxon Corp.)

BUYING REAL ESTATE

There are two ways to buy real estate. You can negotiate the terms, or get the property through condemnation proceedings.

Negotiation

When *negotiating* a purchase, the buyer and seller discuss the terms of sale. The terms of sale include:
- The selling price.
- Possession date (the day the owner gets the deed).
- What remains on the property.

For example, a parcel (piece) of land may sell for $20,000. The buyer gets possession on January 1, 20____. The seller leaves each feature on the land except 27 marked trees. The seller must remove the trees before the date of possession.

People prefer to negotiate a sale. Both parties (buyer and seller) are more apt to be happy.

Condemnation

Public projects and large private utility projects require more land than most private projects. A highway, reservoir, pipeline, railroad, airport, and power transmission line are examples. The goal is to select sites that best serve the public.

Negotiation is tried first. *Condemnation proceedings* are used when an owner will not sell or asks a price that is too high. The state has the "power of eminent domain." This is the right to take or condemn property for a public purpose. The decision is made in court. The court will set a fair value for the land. This process stops a single person from halting the project.

Describing Real Estate

There are millions of parcels of land. A system of lasting records is used. These records show the boundary (outside) lines for each piece of land. The records are kept in the Recorder's Office at the county courthouse. The records include present and past owners. Dates of purchase and transfers of title are all there. *Plats* (maps) show the location of the land.

Legal descriptions specify pieces of land in words. There are three methods used to write legal descriptions. One, the *lot number,* refers you to the subdivision (section of the city) name, and the location of the plat, Fig. 5-8.

Lot 37 of Gray's subdivision as per plat recorded in Plat Book (C) page 37, of the public records of Calhoun County, Iowa.

Fig. 5-8. A legal description by lot number, subdivision, and plat location.

The second method is called a description by metes and bounds, Fig. 5-9. **Metes and bounds** is used to describe land with an odd-shaped boundary. This method starts at a known point. Then the direction and distance to the next point is stated for each side. This method is used until you return to the starting point.

Beginning at a concrete and bronze monument located 27.3 feet south and 47.6 feet west of the center lines of Jefferson Avenue and 15th Street in the town of Lohrville, County of Polk, State of Nebraska.

South 10°17' east along a westerly line of Jefferson Avenue for a distance 125.4 feet to a chiseled X on a manhole rim. Then south 74°54' west for a distance of 97.6 feet to a steel pin, thence;

All bearings being referred to a true meridian; the tract containing a calculated area of 3.89 acres, more or less: and being shown on the plat drawn by Dennis Jorden, Registered Land Surveyor, which is attached hereto and made a part hereof.

Fig. 5-9. A legal description by metes and bounds.

The final method is based on **meridians** (north and south lines) and **base lines** (east and west lines). See Fig. 5-10. The areas between the meridians and base lines are divided into townships. A *township* has 36 square miles (6 miles on each side). The township is divided into 36 *sections* that are one mile square (640 acres). Sections are divided into *quarters* (160 acres). Quarters are further divided into four parts (40 acres). The 40 acre plots are

Southeast quarter of Section 7, Township 3 South, Range 4 West of the Fourth Principle Meridian, County of Harris, State of Missouri, containing 160 acres, more or less, according to the United States Survey.

Fig. 5-10. A legal description using meridians (north and south lines) and base lines (east and west lines).

made into subdivisions. A *subdivision* is a group of lots where homes are built.

The buyer wants a good title to the land. A **good title** means that a legal transfer has been made with no lien against it. A **lien** states that a previous owner still owes money to someone. The money was used for a building or to improve the land.

Paying for Real Estate

A simple way to buy land is to pay cash. The buyer pays the seller the full amount of the terms at the time of the sale.

The purchase is financed when the buyer wishes to spread the payments out over a period of time. Land contracts and mortgages are used when a purchase is being financed.

At the closing, the land changes owners. The seller, buyer, lender, and their attorneys meet. At the meeting, the buyer pays the seller for the land. The buyer may pay cash, sign a land contract, or have a mortgage.

A land contract is a written agreement. The agreement is between the buyer and the seller. Most often there is a down payment. The **down payment** is a portion of the total selling price. The rest is paid for later. The payment may be paid in a lump sum (all at once) on a stated day.

A second way is to make payments. They pay off the loan (principal) and pay for the use of the money (interest). A payment is a set amount. The time between payments and the pay off date are stated in the terms of the contract. For example, with a land contract, a missed payment often means loss of the entire investment.

The terms describe the sale amount and interest. They also describe what can and cannot be done with the land.

More About Mortgages

Mortgages are made by firms who lend money to buyers. A buyer visits with a loan counselor. The loan counselor helps the borrower complete all forms. Help is available to prepare for meeting with a loan counselor, Fig. 5-11. The use of the money is described in detail. The loan committee reviews the application. See Fig. 5-12. In business ventures, the buyer may present the project to the committee.

If the loan is approved, mortgage papers are drawn up and signed. **Mortgages** state the plan for paying off the loan. The payment, interest, pay periods, and payment method are described. See Fig. 5-13.

Cash Budget

(For three months, ending March 31, 20_____)

	January		February		March	
	Budget	Actual	Budget	Actual	Budget	Actual
Expected Cash Receipts:						
1. Cash sales						
2. Collections on accounts receivable						
3. Other income						
4. Total cash receipts						
Expected Cash Payments						
5. Raw materials						
6. Payroll						
7. Other factory expenses (including maintenance)						
8. Advertising						
9. Selling expense						
10. Administrative expense (including salary of owner-manager)						
11. New plant and equipment						
12. Other payments (taxes, including estimated income tax; repayment of loans; interest; etc.)						
13. Total cash payments						
14. Expected Cash Balance at beginning of the month						
15. Cash increase or decrease (item 4 minus item 13)						
16. Expected cash balance at end of month (item 14 plus item 15)						
17. Desired working cash balance						
18. Short-term loans needed (item 17 minus item 16, if item 17 is larger)						
19. Cash available for dividends, capital cash expenditures, and/or short investments (item 16 minus item 17, if item 16 is larger than item 17)						
Capital Cash:						
20. Cash available (item 19 after deducting dividends, etc.)						
21. Desired capital cash (item 11, new plant equipment)						
22. Long-term loans needed (item 21 less item 20, if item 21 is larger than item 20)						

Fig. 5-11. Calculation of loan amounts needed. See lines 18 and 22. (Small Business Admin.)

Fig. 5-12. The loan committee at a bank reviews the loan application. (Merchant's National Bank)

The lender provides the money to buy the land. In exchange, the lender holds the *deed* (ownership paper) until the lender receives the last mortgage payment.

SUMMARY

Before anyone designs a project, the owner must get a suitable site. The goal is to reduce the per unit cost of the product or service. The next step is to make sure zoning and environmental laws allow the land to be used for the future purpose. A building permit is obtained. Finally the site is bought.

Fig. 5-13. Payments may be made in person or sent through the mail. (Merchant's National Bank)

Negotiations and condemnations are used to set the terms of the sale. Land is paid for by cash, land contract, or mortgage.

KEY WORDS

All the following words have been used in this chapter. Do you know their meaning?

Agricultural
Base lines
Building codes
Building permit
Commercial
Condemnation proceeding
Deed
Down payment
Environmental impact report
Good title
Industrial
Interest
Legal description
Lien
Lot number
Meridians
Metes and bounds
Mortgage
Negotiating
Plan
Plat
Residential
Utilities
Zoning ordinances

TEST YOUR KNOWLEDGE

Write your answers on a separate sheet of paper. Do not write in this book.

1. Two steps an owner or initiator follows to obtain a building site are to _____ the site, and buy the site.
2. To calculate the per unit cost for a construction project, the cost of construction is first _____ the cost of operation.
   a. subtracted from
   b. multiplied by
   c. divided by
   d. added to
3. Building codes keep structures in agreement with _____ ordinances.
4. Land zoned for commercial structures allows the buildings to be used for:
   a. Heavy industry.
   b. Railroad cargo handling.
   c. Supermarkets and consumer outlets.
   d. Farming activity.
5. The building inspector grants a building permit after approving the _____ (written description) for the project.
6. You can buy real estate by _____ the terms or by condemnation proceedings.
7. A legal description of land by _____ and bounds is used for odd-shaped land parcels.
8. A notice of a debt still existing for a property is called a _____ against that property.
9. A mortgage is paid off (with a lump sum/by regular payments).
10. A missed payment on a land contract can mean loss of the entire investment. True or false?

ACTIVITIES

1. What is the zoning for the area where you live?
2. Find out if you can start a barbershop or hair salon in your neighborhood. If not, find out why. If you could get the zoning changed, how would you do it? Write about what you learn.
3. Ask a grocery store owner near you why the business selected the site it did for the store.

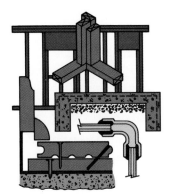

Chapter 6

SURVEYING THE SITE

After studying this chapter, you will be able to:
☐ List some jobs a surveyor does.
☐ List job titles in a survey party.
☐ Explain how to use a level, a steel tape, and a transit.
☐ Explain the need for a soil survey.

The *architects/engineers (A/Es)* of a construction project must know about the features of the site. They must know where the site is located. Roadway engineers need to know the shape of the surface and how the land is used. Surveys describe obstacles that are above and below the surface. For marine projects, facts about the shoreline and the soil under the water are gathered. The kind of soil and rock on the site determines the:

• Route of a roadway.
• Foundation (support) of a building.
• Construction methods used to build the project.

The route of a roadway is designed to avoid obstacles that add to its cost. Buildings placed on rock have a different foundation than if placed on clay. The methods used to build roads in swamps differ from those used on dry land. The A/Es learn about the features from surface surveys and soil surveys.

WHO SURVEYS THE SITE?

The people who survey a site may work for the owner. Surveyors for most public projects are from a company that does special kinds of surveying. These people are known as consultants. Refer to the diagram in Fig. 6-1.

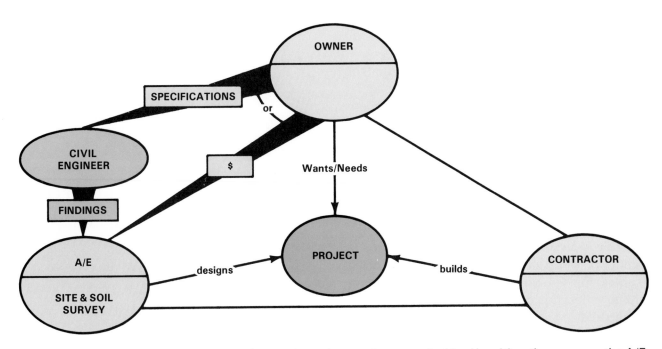

Fig. 6-1. The A/E can do the soil and surface survey. A consultant may be hired by either the owner or the A/E.

Fig. 6-2. Surveying through wooded areas and over rough land takes a large survey party. (Ingersoll-Rand Co.)

A group of people who do survey work is called a *survey party*. The size of the survey party depends on the size of the project. The party size usually varies from two to eight people. Sometimes it is larger. A large survey party was used to survey for the highway in Fig. 6-2.

Members of a survey party do different kinds of work. On large projects, party members may be hired to do just one job. One person may do more than one job on a smaller project. The party chief plans and directs the work.

The *recorder* takes notes on the survey findings. The notes are written in field notebooks. The survey instruments are set up and used by the instrument operator, Fig. 6-3. Survey instruments are tools that help people measure angles and lengths.

The *rod person* handles the *Philadelphia rod* (a painted pole with markings on it) on the point being marked, Fig. 6-4. The points being marked have an unknown location or height, or the points need to be checked.

A survey party may work in an area where there are a lot of weeds and brush. An axe person is

Fig. 6-3. Using level-transit. Hand signals are used by the instrument operator to let the person holding the rod know if it is held right. (David White Instruments)

Fig. 6-4. This rod person is holding a Philadelphia rod. Laser detectors can be mounted on leveling rods to check differences in elevation of setting grade. (The Lietz Co.)

needed. The axe person clears brush away so that the rod may be seen from the instrument. These people may build towers or signals to get above the trees in a wooded area.

People who survey use math, physics, and astronomy (knowledge about the stars) to do their work. Land surveyors need to know about legal matters that relate to owning land. Surveyors must know how to make neat and accurate drawings. Being thorough and working in a planned way is important. This way of working helps a surveyor to find sources of error and correct them in the final results.

WHAT DO SURVEYORS USE?

Survey work must be exact. Using instruments properly is the only way to make accurate surveys.

Two different groups of equipment are used. The groups are surface survey equipment and soil survey equipment.

Surface Survey Equipment

A *transit*, Fig. 6-5, is used in the field. It measures vertical (up and down) and horizontal (flat) angles. A transit is a telescope. The telescope has cross wires for taking vertical and horizontal readings.

The *level* is shown in Fig. 6-6. Like the transit, it has a telescope with crosswires and bubble tube. It has a marked circle to measure horizontal angles only.

Newer equipment is now used to level. Some surveyors use a device that sends out a rotating *laser* (light) beam. After setting up and getting the laser running, the worker can go to the points that need to be checked. The worker uses a special tool that "sees" the laser beam and checks height. Some de-

Fig. 6-5. A transit will measure both horizontal (left and right) angles and vertical (up and down) angles. (David White Instruments)

Fig. 6-6. A level measures only horizontal (left and right) angles. (TOPCON, Inc.)

vices even have an automatic leveling feature. If the tool is bumped while in use, it will level itself.

Steel tapes, Fig. 6-7, are used to measure horizontal distances. An *electronic distance meter (EDM),* Fig. 6-8, is used where the terrain is rough or when distances are long. Shorter distances are best measured with the tape.

A *plane table* is used for making drawings in the field while making the survey. Fig. 6-9 shows one in use. It consists of a drawing board and a tripod (three legged support).

An *alidade* is used to make measurements on the plane table. The edge of an alidade is a rule for drawing on the paper on the plane table.

A *plumb bob* is used to locate a point directly under another point. Philadelphia rods of metal or wood are used in reading elevations (heights). Look again at Fig. 6-3.

Hand-held *sighting levels* are used to find a line of sight. The stadia wires, level bubble, and crosswires are all seen at the same time in sighting levels, Fig. 6-10. Usually the intent is to adjust it until the

Fig. 6-8. This electronic instrument (total station) combines distance and angle measurement into a single unit. Survey data is stored electronically for later use. (The Lietz Co.)

Fig. 6-7. Steel tapes may change in length when it gets warmer or colder. A special metal is used to reduce the effect of heat. (The Stanley Works)

Fig. 6-9. A plane table is used with an alidade to make drawings in the field. (Ball State University Photo Service)

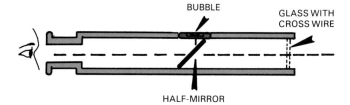

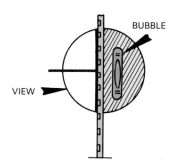

Fig. 6-10. The hand-held level uses a half-mirror. A person can see the bubble and the rod at the same time.

Fig. 6-11. Larger samples from deep below the surface are taken with power drills. (Western Technologies, Inc.)

wire cuts the bubble image in half. Then note where the wire falls on the rod being viewed. Sighting levels are used to make rough measurements.

Soil Survey Equipment

Soil samples are taken with *augers*. The tools are screwed into the soil and a sample of soil is removed. The tool used depends on the kind of soil and the kind of sample needed. Some are turned by hand. Motor driven tools are used for large samples from deep holes in hard soil, Fig. 6-11.

Geophysical methods are used for large areas. *Geophysical methods* measure characteristics of soil and rock without taking samples. The *resistivity method* uses electricity. It is a method that helps plan foundations. The test measures how well the soil conducts electricity. Changes in the flow of electricity due to water mean the soil is different. The water content is important for planning drainage needs.

Seismic methods are based on the fact that sound travels faster in denser material. Refer to Fig. 6-12. A sledge hammer is used to create a sound wave. Sensitive electronic equipment hears the shock wave. The equipment shows the density of the material on a scale.

In the past, people were careless in disposing of toxic (poison) materials. Some didn't know the materials were dangerous. Others just didn't care. We don't know where these materials were dumped. When hazardous materials are found, they need to be removed before construction begins, Fig. 6-12.

Fig. 6-12. Technicians are taking soil samples from a site likely to have hazardous material. (Western Technologies, Inc.)

HOW ARE SURVEYS DONE?

Most surface and soil surveys are *resurveys.* This means that a survey of the area was done before, and is now being done again. The United States Geological Survey has mapped most of the country. Today's surveys are done to check known points or to get more details.

A survey is done by the steps listed in Fig. 6-13. The surveyor must:

- *Know purpose* — Find out what is to be learned from the survey.
- *Do research* — Look for information that already exists.
- *Do field work* — Find new or aquire more detailed information.
- *Do plotting* — Record the findings.
- *Make drawings* — Prepare drawings in a form designers can use.

Purpose

In order to make the survey, the chief must know the purpose. The party chief receives the information from the owner or A/E.

Research

The cost of a surface or soil survey can be reduced greatly if the consultants use surveys already available. The process of looking for surveys is called *research.* Soil and bedrock facts can be found in local, state, and federal offices.

Public utilities keep surveys. They show elevations and the location of land, utility systems, and structures. Railroads have records of carefully made surveys.

Field Work

All future land survey work begins from known points found by research. Facts about the features of a site are gotten by the survey party in *field work.* The first visit to the site is for reconnaissance (to look the area over). This visit makes planning the survey easier.

In field work, known markers are found first. Checking the known markers' accuracy is done next. Lastly, new markers are set.

Soil surveyors look at the site. They test soil in the field. These are called *field tests.* Also, samples are taken and tests on the samples are done in a laboratory. These are called *laboratory tests.*

Plotting

All points and measurements are plotted in field notebooks. Notebooks used by some early survey parties are found in some city and county surveyor's offices. Recording data (making notes) is done with care. Notebooks or electronic devices are used. Fewer errors and less time is needed with electronic devices.

Drawing

The facts found in the field are used to make drawings of the site. These drawings describe the heights or shape of the site.

The information that is gotten in the field needs to be made more useful. Math is used to compute unknown values.

The useful findings are given to the A/Es and they make drawings of the site. These drawings help in the design of the project.

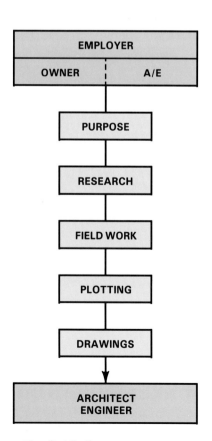

Fig. 6-13. Survey sequence.

HOW IS THE EARTH'S SURFACE DESCRIBED?

A *surface survey* locates points on or near the earth's surface. The points describe boundaries, the shape of the earth's surface, and where to place structures. The *subsurface survey* finds out what the soil is like under the surface of the ground.

You start your study with surface surveys. These include the control survey and also the land and topographical surveys.

Control Surveys

Most of the country has been surveyed. Selected points have been placed and marked with monuments. A *monument* is a fixed marker. It is made of brass or other metal that withstands weather. Monuments are attached to a solid foundation (base). Rock or concrete placed deep in the soil is often used. Fig. 6-14 shows how one kind of monument is made.

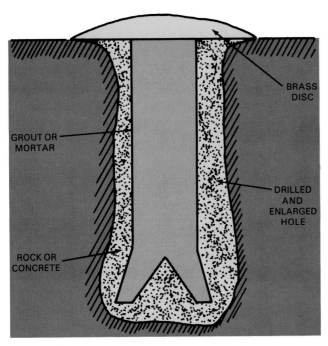

Fig. 6-14. Monuments (fixed markers) are made in many ways. This method is used in solid rock or concrete.

Both horizontal positions and elevations are marked. This system of monuments is known as the *control survey.* The monuments serve as starting points for new surveys or for checking other surveys. A horizontal control locates points and distances that are far apart. A vertical control gives elevations.

Land Surveys

A *land survey* is made to find out the location and size of a parcel of land. The survey party measures the length, direction, and placement of boundary lines. Next, they compute the area the boundary lines enclose.

Boundary lines may define an area. Fig. 6-15 displays one method that uses a series of lines that meet at angles to surround the area.

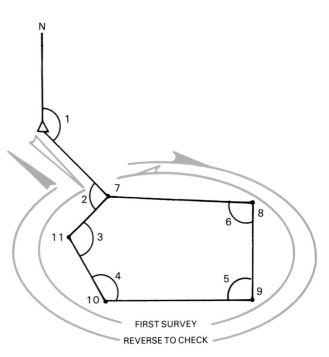

Fig. 6-15. Traversing to set boundary lines. The first survey is checked by reversing the steps.

A second method is to establish a *survey line.* The area includes the land on each side for a stated distance. This method is used for long narrow projects. Roadways, pipelines, transmission lines, and canals are examples, Fig. 6-16.

When all of the points cannot be seen, surveyors build a tower and signal. These structures make it possible to see over trees and rocks. A technique called offsetting, Fig. 6-17, is used to avoid things in the line of sight. When *offsetting,* you take sightings off to the side of the actual boundary to avoid the obstacle.

Fig. 6-16. A pipeline survey follows a center line. Zig-zag allows for expansion and contraction. (El Paso Co.)

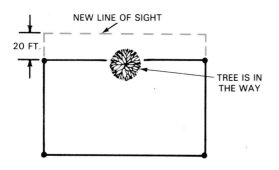

Fig. 6-17. Offsetting is a way to get around things in the line of sight.

Topographical Surveys

A *topographical survey* is done to describe heights of the surface of the land. The shape of the surface and the features on and below the surface are shown in Fig. 6-18.

The structures on the site, and where they are, will affect the design of the new project. The location of water lines, sewers, telephone lines, and electricity lines need to be known. All of these features are a part of the topographic survey.

The survey party finds out who owns the adjoining land and how it is used. These facts are used in buying the site. The location of trees, roadways, and drainage ditches must all be known before the new project is designed. A topological drawing locates and describes each feature.

A/Es need to know the shape of the building site. Two leveling methods are used to measure the shape of the surface. The methods are differential and stadia.

Differential leveling uses the horizontal line of sight as a reference. See Fig. 6-19. The line of sight is kept level with a spirit bubble.

The crosswires in the level make it possible to measure heights accurately. See Fig. 6-20. The surveyor's rod measures the heights. The horizontal wire is used to make elevation readings. The vertical wire shows if the rod is held straight up and down. Elevation between two points is found through subtraction.

The *stadia leveling* technique is faster but is less accurate. See Fig. 6-21. The line of sight does not have to be in the horizontal plane. Sightings are over longer distances. Therefore, fewer sightings need to be taken. Very hilly land is leveled by using the stadia leveling method.

Surveyors can compute desired measurements after taking just a few simple readings. This explains the time saved when compared to the differential method.

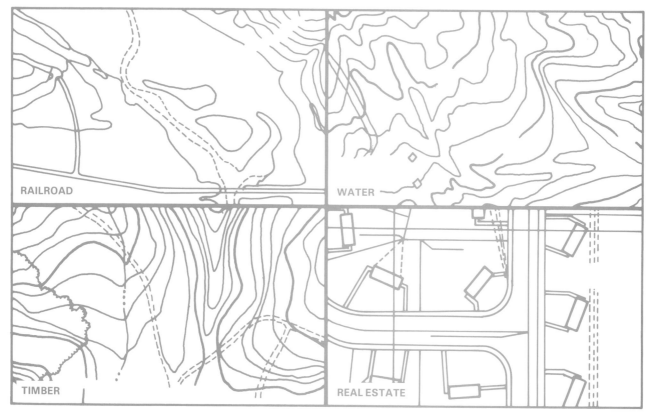

Fig. 6-18. Features above and below ground are shown on this topographical drawing.

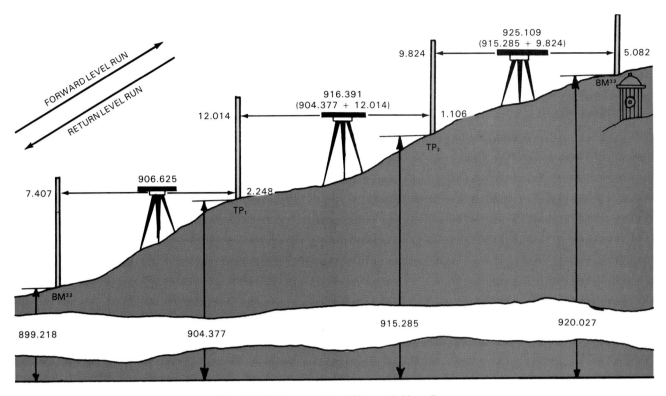

Fig. 6-19. How to do differential leveling.

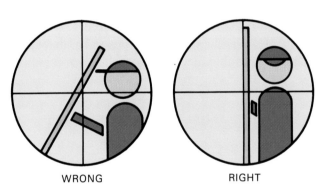

Fig. 6-20. The crosswires are used to see if the rod is straight up and down and to read heights.

Fig. 6-21. In stadia leveling, surveyors measure distances and angles. They use math to figure unknown distances.

WHAT MAKES UP THE EARTH'S CRUST?

The earth's crust consists of rock and soil. Refer to the chart of soil types in Fig. 6-22. The surface of the earth ranges from solid rock to fine clay. The Rocky Mountains have many areas where solid bedrock is visible. The midwestern states are covered with different kinds of clay, silt, and sandy soils.

ROCK

EARTH TYPE	SIZE
Bedrock	Solid
Boulders	8'' to several feet across
Cobbles	4'' to 8'' across

SOIL

EARTH TYPE	SIZE
Gravel	Peas to 4'' in diameter
Sand	Salt to pea size
Silt	Baking powder to salt
Clay	Baby powder to baking powder

Fig. 6-22. The make-up of the earth's crust.

Characteristics of Soil

Soil has many characteristics. The characteristics are the result of the size and shape of the particles. The open space, called a *void,* between the particles varies as the size of the particles varies. The larger the particles the larger the voids. Soil with large voids is called loose soil. Dense soil has small grains and small voids.

The mixture of soil particles may vary greatly from one area to another. This is a concern on sites used for projects like railroads. How soil differs can be seen by walking in the plowed field of a farm. Samples are taken, Fig. 6-23, to learn about the soil.

The water in soil affects its characteristics. Water can come from drainage. Drainage water is present in most soil.

Another form of water in soil is underground water. This water forms rivers and lakes under the surface of the ground. Most of the water is in motion. Unlike water on the surface, it moves slowly through the soil. Underground water fills voids between soil particles.

In some areas of the country, water under the ground has made the rock dissolve. Great quantities of rock are gone. Where the rock is gone, caves are formed. When the rock or soil over the caves

gets weak, it gives way. The soil above drops into the cave and forms a *sink hole* on the surface.

Some clay will *slake* (get soft) when it gets wet. Clay that loses its strength when it gets wet does not make a good base for a structure. In this case, special foundations need to be designed. See Fig. 6-24. Wood pilings (poles) are driven deep into the soil when the soil is weak. These vertical pilings act as a foundation for further work.

Most soil swells when water is added to it. When it dries it shrinks. Dense soils will crack when they dry. Loose soils are weak, and cracks do not form.

Wet soil expands when it freezes. When this happens the surface rises. Structures sitting on the frozen surface will be raised.

The colder the weather the deeper the soil freezes. The greatest depth that the soil freezes is called the *frost line.*

SUMMARY

Surveys provide data about the building site. Land surveys locate the site. Topographical surveys describe the surface and what is under the soil. The nature of the soil is found in soil surveys. A survey party measures the shape of the ground under water in a hydrographic survey. Before the oil port in Fig.

Fig. 6-23. Soil on an Arizona ranch is being tested. The mixture of soil grains changes. The depth to the bedrock varies from place to place. (Western Technologies, Inc.)

Fig. 6-24. This foundation is called a friction pile type. The soil is deep and gets soft when it is wet. (Koppers Co., Inc.)

6-25 was built, several kinds of surveys were made. Can you list the kinds?

Survey parties work outside in groups of two or more people. The chief works inside when doing research. Some physical labor is done on very rough terrain. See Fig. 6-26. It is hard work, but the outdoor areas are sometimes pleasing to see.

Fig. 6-26. Soil surveys are not easy to make. People in this survey party probably like to work outside. (Raymond International, Inc.)

Fig. 6-25. What kind of surveys were needed before this offshore natural gas port was built? (Arabian American Oil Co.)

KEY WORDS

All the following words have been used in this chapter. Do you know their meaning?

Alidade
Architect/engineers (A/Es)
Auger
Boundary line
Control survey
Differential leveling
Electronic distance meter (EMD)
Field test
Field work
Frost line
Geophysical method
Laboratory test
Land survey
Laser
Level
Monument
Offsetting
Philadelphia rod
Plane table
Plumb bob
Recorder
Research
Resistivity method
Resurvey
Rod person
Seismic method
Sighting level
Sink hole
Slake
Stadia leveling
Steel tape
Subsurface survey
Surface survey
Survey line
Survey party
Topographical survey
Transit
Void

TEST YOUR KNOWLEDGE

Write your answers on a separate sheet of paper. Do not write in this book.

1. A (level/transit) measures both vertical and horizontal angles.
2. EDM is an abbreviation of what?
3. A sighting level is adjusted until the cross wires cut the _____ image in half.

4. The edge of a(n) _____ is a rule for drawing on the paper on a plane table.
   a. sighting level
   b. alidade
   c. transit
   d. plumb bob
   e. level-transit
5. The principle of a _____ soil density test is that sound travels faster in a denser material.
6. What is a resurvey?
7. Arrange in their proper order the following steps in surveying.
   a. Do field work.
   b. Make drawings.
   c. Do research.
   d. Know purpose.
   e. Do plotting.
8. An area for a long narrow project like a pipeline is established by a _____ line.

Matching questions: On a separate sheet of paper, match the definition in the left-hand column with the correct term in the right-hand column.

9. _____ Locates points on or near the earth's surface.
10. _____ Lists soil types at several points.
11. _____ An existing survey with permanent marking objects.
12. _____ Permanent marking objects which withstand weather.
13. _____ Done before computing area of land parcel.
14. _____ Avoids things blocking the line of sight.
15. _____ Describes heights of the surface of the land.
16. _____ Line of sight not always in horizontal plane.

a. Land survey.
b. Topographical survey.
c. Control survey.
d. Offsetting.
e. Subsurface survey.
f. Stadia leveling.
g. Surface survey.
h. Monuments.

ACTIVITIES

1. Go to your principal's office. Ask if you can see the set of drawings for your school. Find the site plan. What does it show?
2. Go to the water treatment or waste water treatment plant. Ask for the city topographical

map. What does it show? Ask the person in charge what information they use on the map.

3. Find a legal description of the land where you live. You can find it on the legal papers your parents got when they bought the land. If they do not own the land, the legal description can be found at the county courthouse in the recorder's office.

Tools needed for the site are often housed in temporary shelters. (Ingersoll-Rand Co.)

Section 3
DESIGNING AND PLANNING THE PROJECT

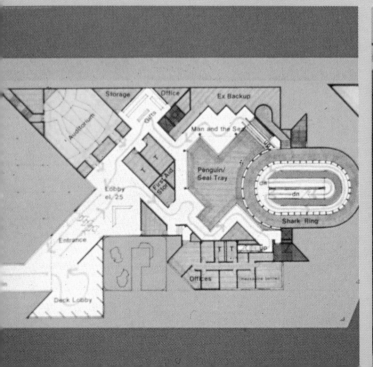

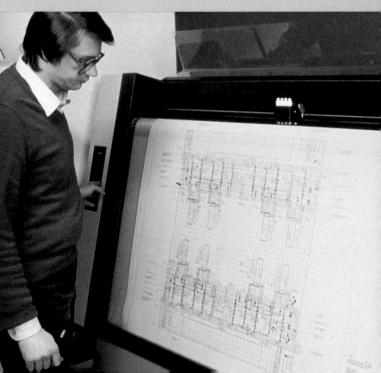

Chapter 7

DESIGNING THE PROJECT

After studying this chapter, you will be able to:
☐ Contrast an architect's work with the work of an engineer.
☐ List the five factors designers must consider.
☐ Discuss if it is good or bad that problems become limitations.
☐ Describe some ways to select and refine ideas.

Designing is an activity which finds solutions to a problem. Suppose that you are going to build a birdhouse. You would need to have an idea how to build it. Then you buy and cut lumber.

The birdhouse must meet certain conditions. It must be light enough to hang from a tree branch and big enough to house the size bird you wish to attract. The materials must withstand all types of weather conditions. Yet, it must be attractive and inexpensive. Finally, it must be buildable with the tools you have available.

Designing of large structures such as skyscrapers, bridges, dams, and roadways requires the same steps as you would follow in designing the birdhouse. Of course, the larger the project, the more complex the designing process becomes. See Fig. 7-1.

Designing large projects takes a long time. The time is well spent because construction projects are used for a long time. It is hard to change them once they are built.

Fig. 7-1. Designing large buildings is complex. (Cambridge Seven Associates, Inc.)

In most cases, the design of the construction project begins before the owner selects the site. The site for the project should fit into the master plan. If the site is chosen first, the master plan must be made to fit the site.

WHAT IS DESIGNING?

Creative designers must be able to see a problem and find the best way to solve it. Solutions to problems are usually unique (one of a kind). They are unique because each problem differs slightly.

As you travel on a highway, many bridges look alike, Fig. 7-2. If the bridges are studied more

Fig. 7-2. All bridges may look alike, but they are not. Each is unique.

closely, you will find that each is unique. The span, grade, and foundations differ for each bridge. The designers used standard materials and methods, but used them in different ways for each bridge. A unique design was the result.

People can build on knowledge. We can save knowledge in our brains, in books, and in many other ways. One of the best ways is in computers. What we learn, we record. We can then retrieve (get) it back again.

For example, the longest suspension bridge in the world has a main span of 4,626 feet. The principles used to build this bridge are the same as the principles early people used. The principles were built upon. New materials were used. A longer bridge that supports more weight was the result, Fig. 7-3.

Computers have helped a lot in designing structures. They store information in an organized form. It is easy to retrieve and use the information. Best of all, computers do things quickly. Fig. 7-4 shows a computer aiding in drafting.

Fig. 7-3. A/E build on knowledge that already exists. This bridge is an example. (HNTB Engineers)

Fig. 7-4. Much of this engineer's time is saved by using computers to produce drawings. This field of work is called computer-aided drafting (CAD). (Le Messurier Consultants)

Architects and engineers (A/Es) take knowledge and build on it. They use what has been learned to build taller buildings, larger dams, longer bridges, and bigger harbors.

WHO DESIGNS CONSTRUCTION PROJECTS?

The *project designer* is hired by the owner. Who designs a project depends upon what needs to be built. Small projects at home are designed by people like you and your parents. Fathers and mothers can design complete homes, and additions to their homes, Fig. 7-5.

A/Es are professionals who design projects. A **professional** is a trained person who is paid for a service. They are hired by the project owner. The person who pays for the service is known as the **client.** A client is an owner, represents an owner, or is a civic leader, Fig. 7-6.

Fig. 7-5. This family designed their own home. Now they are choosing the cabinets to put in it. (Pella/Rolscreen Co.)

Architects are hired to design buildings and do master planning. A **master planner** sets the main theme for a large complex project. See Fig. 7-7. A shopping center is made up of many stores, walking malls, parking lots, and roadways. The master planner sees that all parts of the project fit.

Building construction projects are designed by architects. Housing and commercial and public buildings are examples. For larger buildings, engineers design the foundation, framework, and mechanical systems. **Architects** consider the looks and function of the building and select materials. **Engineers** consider the strength, utilities, and drainage, Fig. 7-8. Engineers are chief designers of dams, pipelines, roadways, water control systems, and utility systems.

WHAT ARE THE CONCERNS OF THE DESIGNER?

A/Es are always aware of five factors, Fig. 7-9. These factors are:
- **Function** - Will the project do what it is supposed to do?
- **Appearance** - Does the project look good and does it blend with the area around it?
- **Cost** - Will the project be within the budget?
- **Strength** - Will the project withstand the forces of nature?
- **Materials** - Which materials will be the best for the purpose?

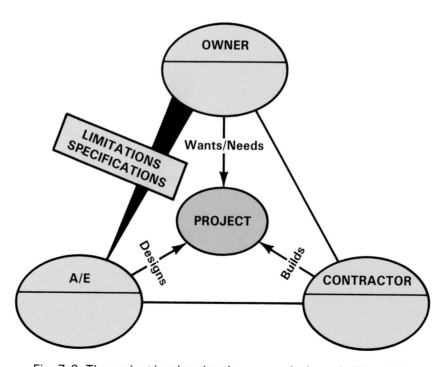

Fig. 7-6. The project is related to the owner-designer-builder triad.

The problem is to design a prototype one-bay hangar. The hangar is to have:
1. Space to enclose either a Douglas PC-8 or Boeing 727 aircraft, (18,000 sq. ft.).
2. Support space for line maintenance (14,000 sq. ft.).
3. Cargo facilities for air freight (18,000 sq. ft.).
   The one-bay hangar is to be designed so that it can be built singly or in combinations.
   The cost should be approximatley $2,000,000.

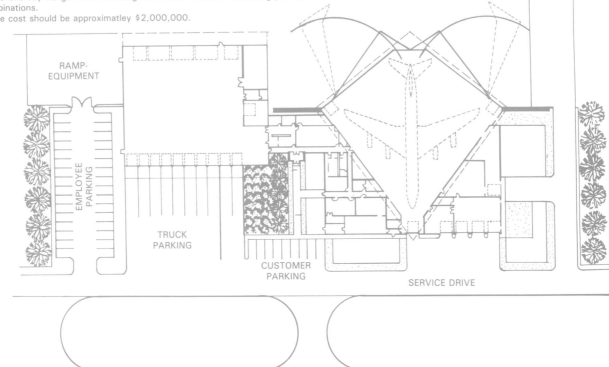

Fig. 7-7. Architects did the master plan. An engineering firm designed the special roof. A second engineering firm designed the mechanical systems. (Miller-Dunwiddie Architects, Inc.)

Function

Function is the most important concern for a structure. A bridge must have enough lanes to handle the traffic. It must be high enough so ships can pass under it.

Power generators must be efficient and safe to operate. Designers must make sure that the project will work the way it should. See Fig. 7-10.

Appearance

The appearance of a project should be right for the setting. See Fig. 7-11. A tall building should add to the effect of its neighbors. They will stand out on the city's skyline. Many people see them. Bank buildings are made to look secure. They should convey the thought that you can rely on the people inside.

Cost

It is vital that the A/E controls the cost of the project. Most often the A/E of a project is given a *budget*. The owner wants to get the most for the money. To the A/E for a retail store, it means the most square feet of floor space for a set price. To manufacturers, the goal is to keep the per unit cost of the product low. This means that construction and operating costs must be held down.

Strength

Engineers use many clever ways to give strength to a structure. See Fig. 7-12. A project must be strong enough to withstand the forces of nature. A dam must withstand great pressure from the water. Skyscrapers must withstand strong winds and a lot of weight. Structures in certain parts of the country must withstand earthquakes. Pipelines are designed to withstand pressure from the inside.

Materials

An A/E selects the best material for the purpose. Materials serve a function. Marble is used to show richness and to last a long time. Steel gives strength. Glass gives a sense of space and lets in light. Con-

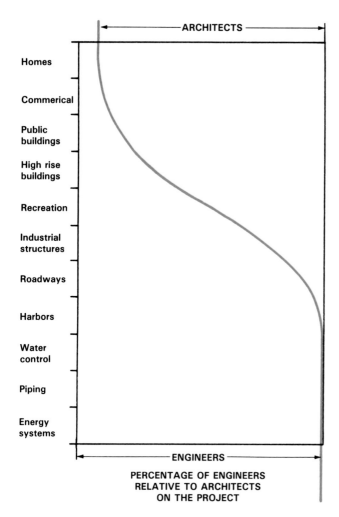

Fig. 7-10. The trains and tracks must work in this mass transit project. The station and trains must also be able to handle the people. (HNTB Engineers)

Fig. 7-8. Who designs what projects? This chart will help.

crete provides a strong mass. A/Es must know about materials and their characteristics. By learning about materials they can select the best ones for the project.

No single factor stands alone. They all relate to each other. The A/E looks at each factor in light of all the others. A material may look right. If it costs

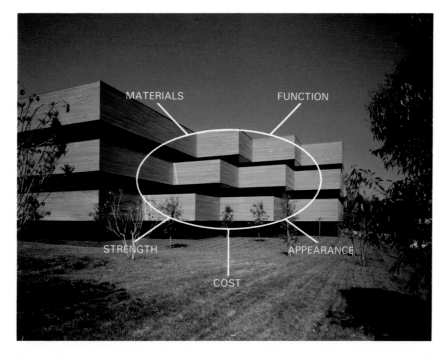

Fig. 7-9. Can you state how the five concerns influenced this building? (California Redwood Assoc.)

Fig. 7-11. Was the architect successful in making this building attractive? (Graham Gund Architects)

Fig. 7-12. The triangle is a strong shape. Here it is used to support a dome. (TEMCOR)

too much and is not strong enough, it is a poor choice.

A strong and low cost framework is not a good choice if it obstructs the floor plan. The A/E must keep all factors in mind at all times. Value placed on each factor differs as projects differ.

THE DESIGN PROCESS

Designing starts in the early stages of the project. It continues throughout the life of the structure. The goal is to get the most value for the cost. A/Es try to find and remove items that add cost without helping the function. The further designers get into the project the less impact (effect) they have on the project. See Fig. 7-13.

The A/E approaches a problem in many ways. Good A/Es use a planned approach to solving problems. Recall the birdhouse. It was a simple problem. The approach used for large projects requires much more of the same kind of planning. The design of a dam and lake, for example, takes more planning than the design of a farm pond. More planning is needed because large projects are more complex and more people are involved. Many A/Es use the steps shown in Fig. 7-14.

Step I: Predesign – Identify the Problem

In the *preddesign step,* the problem is identified. Predesign work defines the project. It usually consists of:

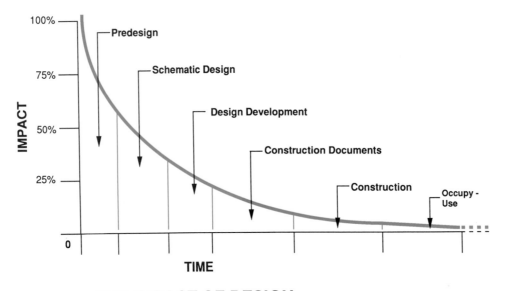

THE IMPACT OF DESIGN

Fig. 7-13. Designing goes on throughout the life of a structure. Designers affect the project most in the early stages.

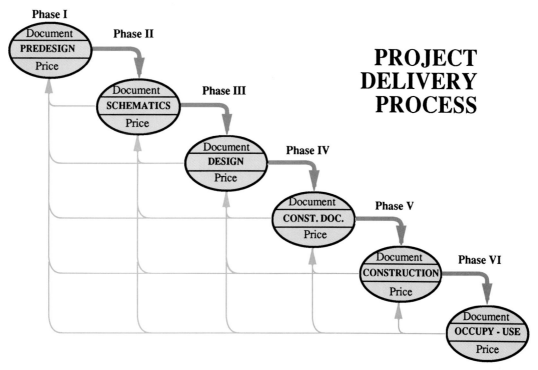

Fig. 7-14. An organized approach is used to define a problem and find a solution.

- *Program* - What goals are to be met with the structure?
- *Financial package* - Will the structure earn a profit for the owners?
- *Schedule* - When is the project to be finished?
- *Budget* - How will the money be spent?

Step II: Schematic Design – Generate Ideas

During the *schematic design step,* A/Es generate as many solutions as possible. Little concern is given to details. Every thought that comes to mind is written down. Sketches and notes are used. It is easy to change and improve your ideas. You need only an eraser and pencil.

Brainstorming is a way to encourage everyone to share good ideas. There are three rules to brainstorming. People who use the method:
1. Come up with as many ideas as they can. (The more, the better.)
2. Think up unique ideas. (The wilder, the better.)
3. Avoid judging the ideas. (All thoughts are good.)

It is likely that a lot of new ideas will come out. Perhaps they can be combined to meet the needs of the project.

Step III: Design Development

The *design development step* has two tasks. They are to refine the ideas and to select the best design.

Task 1 – Refine ideas

Two or more of the best ideas are chosen. The A/E refines and improves them in the design development step. Structures are drawn to scale on a site plan. Details and sizes are added to the drawing. As A/Es refine the ideas, they see if the ideas will work, Fig. 7-15. A careful examination is made to see if the project will function in the way that meets the needs of the owner, can be built on schedule, and does not exceed the budget. Refer to Fig. 7-16.

If the A/Es develop more than one idea, they will have a choice. If they compare and combine ideas, better solutions can be found. Each is judged on how well it meets the needs of the design.

Large and complex projects require models. A *model* is a small likeness of a project. Models make it easier to study each design feature. The analysis of an idea may show that it is not a good one. If it does not work, the A/E selects, refines, and analyzes a second choice. Fig. 7-17 is a model of a refinery.

Task 2 – Select the best idea

The next step is to select the best design. The design is selected by the designer or the client. Each idea is shown to the person or group who makes the choice. Models, pictures, slides, drawings, charts, and graphs help to show people the features of each design, Fig. 7-18. Complete drawings may be used for small projects. Drawings for large projects would have less detail. The site, shape, and place-

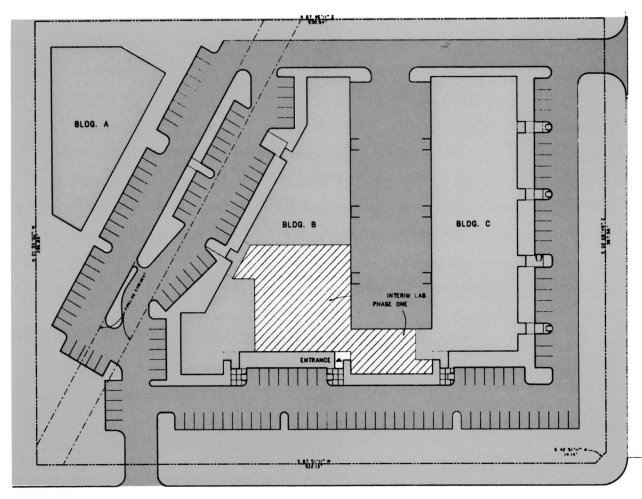

Fig. 7-15. The site plan was only one of the ideas developed. It shows the placement of the building and parking spaces. An interim (temporary) structure is sometimes a part of the plan. (Pierce, Goodwin, and Alexander)

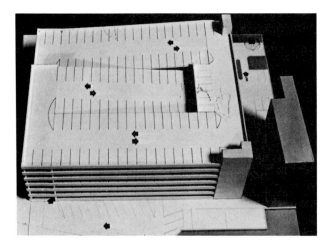

Fig. 7-16. Models make it easy to explain and show parking and traffic patterns. (Ball Memorial Hospital)

ment of the structure would be shown. See Fig. 7-19. Site plans may be made to show buried utility lines or pipes, but details would not be placed on the drawings.

Fig. 7-17. This is a model of a refinery. A model lets the designer see how things fit. It is easier to change the model than a new refinery. (Combustion Engineering, Inc.)

Fig. 7-18. Things people can see are used to present project ideas. A picture (or an object) is worth a thousand words. (DYK Prestressed Tank, Inc.)

When shown the drawing, the client may decide to:
• Accept one design.
• Combine two or more designs into one.
• Reject all designs.

If they reject all designs, the process is stopped, or it starts over again with new ideas.

Step IV: Construction Documents – Implement Design

When the design has been chosen, the A/E prepares construction documents. ***Construction docu-***

ments include the working drawings along with specifications.

Working drawings give the facts needed to build the project, Fig. 7-20. Drawings are used to describe the shape, size, and placement of the parts. Specifications describe in words those things that cannot be shown in drawings. Other documents are added to make up the bidding documents.

Fig. 7-20. The A/E tells the contractor how to build a project with a set of drawings and specifications. (Stran Buildings)

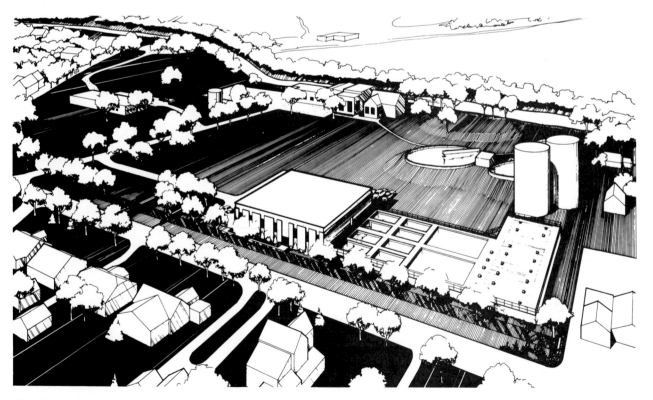

Fig. 7-19. An artist's rendering (drawing) is used to show how structures fit into the surroundings. (Indiana-American Water Co., Inc.)

SUMMARY

The design process begins soon after the decision to build is made. A project designer's goal is to find the best way to meet the specifications for the project.

Architects and engineers design construction projects. They are concerned with the function, looks, cost, strength, and materials of the project. A planned approach is used by those who design projects. The more complex the project, the more important it is to use a planned approach.

KEY WORDS

All of the following words have been used in this chapter. Do you know their meaning?

Appearance
Architect
Brainstorming
Budget
Client
Construction documents
Cost
Design development step
Engineer
Financial package
Function
Master planner
Material
Model
Predesign step
Professional
Program
Project designer
Schedule
Schematic design step
Strength

TEST YOUR KNOWLEDGE

Write your answers on a separate sheet of paper. Do not write in this book.

1. We save knowledge in brains, in books, and in _____.
2. The person who pays for the professional design service is called the _____.
3. The most important purpose of a designed structure is _____.
4. The _____ shape is used when strength and low weight are needed.
   a. hexagon
   b. square
   c. symmetrical
   d. triangle
5. During brainstorming, you avoid _____ any ideas.
6. What is the purpose of a model?
7. An artist's _____ shows how structures fit into the surroundings.
8. List at least three of the five requirements (things the client expects) a designer looks for in a building.
9. For the lowest cost of a walkway above a street, would you choose concrete or steel? Why?

ACTIVITIES

1. Think of a project you would like to build at home. Go through the six steps of the design process. Repeat the process with a friend. Are two heads better than one?
2. Go with a friend or two and talk with an A/E. Talk about how a project is designed. Record your visit any way you want. Report your findings to the class.
3. Can you use the six-step process to earn more money or to plan a trip? Why?

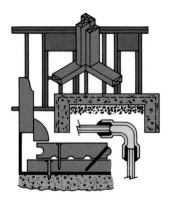

Chapter 8

MAKING WORKING DRAWINGS AND SPECIFICATIONS

After studying this chapter, you will be able to:
☐ Recognize a scale drawing.
☐ Tell how prints of drawings are made.
☐ Name some of the drawing types.
☐ State the purpose of written specifications.

If you want to design a garden shed, drawings can help. You can draw out ideas and save them. It is easy to change your ideas. Walls are easy to move on a drawing. You need only an eraser and pencil.

If you are going to use someone else's idea, this is often provided as a working drawing. The drawing gives you the facts needed to build the project.

In construction projects, designers are seldom the project builders. Working drawings and specifications tell the builder how the A/E wants the project built. See Fig. 8-1. Working drawings show the shape and size of each part and where it goes. The specifications describe in words what materials and methods are to be used. The working drawings and specifications are approved by the owner.

Construction projects are too large to be drawn full-size. Working drawings are drawn to scale. That means they are smaller than the actual project. Each dimension is put on the drawing and it is made easy to read. This saves the workers some time, and they make fewer mistakes.

WORKING DRAWINGS

Working drawings describe the physical details. See Fig. 8-2. Physical details are the shape, size, and placement of parts of the project. The part may be a bridge or a bolt.

Many sets of working drawings are needed. Sets are made for all who are involved in the project. See Fig. 8-3.

The drawings are used to estimate and bid the project. The *estimate* is a careful guess about how much it will cost to build the project. A *bid* is an offer to build the project at a stated price.

Working drawings become a part of the contract when it is signed. Working drawings are the most useful to the contractor and the workers.

Drawings with greater detail are made by the contractor. These drawings are called ***shop drawings***. Shop drawings guide the work of subcontractors, Fig. 8-4.

A Set of Working Drawings

Small projects, such as a house, have all drawings in one set. A site plan, foundation plan, floor plan, elevations, and sections with details are included in the set.

The ***site plan*** shows the horizontal dimensions of the land, Fig. 8-5. The site plan includes the bound-

Fig. 8-1. Graphics generated on computers save a lot of time. Engineering data is fed into the computer. The plotter makes the drawing. (Stran Buildings)

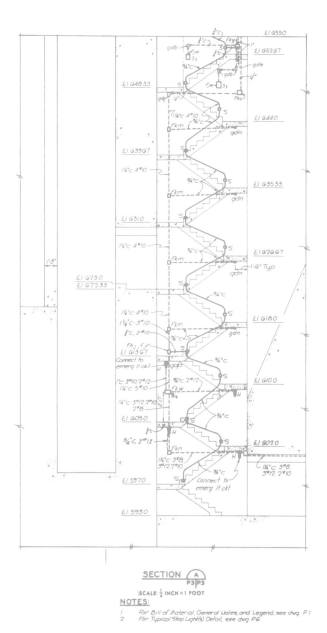

SECTION A
P3 P3

SCALE: ½ INCH = 1 FOOT

NOTES:
1 For Bill of Material, General Notes, and Legend, see dwg P1
2 For Typical Step Light(s) Detail, see dwg P6

Fig. 8-2. Working drawing of a stairwell shows location of conduit fixtures and switches.

ary lines. Vertical controls (bench marks) are placed at each corner.

The placement of structures is found on the site plan. Streets, driveways, utilities, and easements (strips of land you let a utility use) are drawn on the plan. On house plans, the roofline may be on the drawing.

The final shape of the site is shown with contour lines. The placement of trees, shrubs, and fences, and the locations of other landscape features are found on the site plan. If the landscape drawing is complex, a *landscape plan* is added to the set of drawings.

The *footings and foundation plan* shows the building supports, Fig. 8-6. Footings are the base

Fig. 8-3. Special machines are used to make prints. Prints are made from the original drawing. (Lodge-Cottrell/Dresser Industries)

on which the foundation rests. Footings for piers and columns are also shown. Dimensions help show the size, shape, and placement of all footing and foundation parts.

Section views of the footings and foundations are used to show more detail. A section view is a drawing of an imaginary cut through the foundation. These details show exact dimensions and shapes.

The size and placement of steel reinforcement in the concrete is shown in the section view. Concrete is made stronger by placing steel rods in it. These rods are called *rebar* (reinforcing bar).

Fig. 8-4. This electronic track system is for a mailroom in a large corporate office. The workers use shop drawings to guide the work. The shop drawings were made by the subcontractor. (Translogic Corp.)

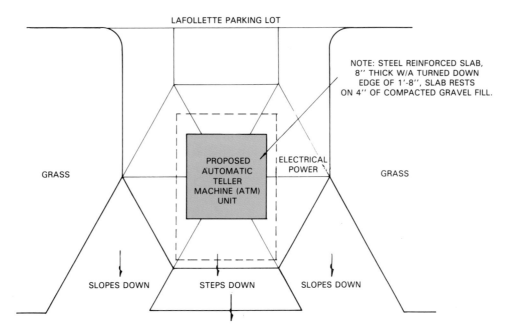

LAFOLLETTE PARKING LOT

GRASS

PROPOSED AUTOMATIC TELLER MACHINE (ATM) UNIT

ELECTRICAL POWER

GRASS

NOTE: STEEL REINFORCED SLAB, 8'' THICK W/A TURNED DOWN EDGE OF 1'-8'', SLAB RESTS ON 4'' OF COMPACTED GRAVEL FILL.

SLOPES DOWN   STEPS DOWN   SLOPES DOWN

Fig. 8-5. Site plans describe the site where the structure is placed. This structure is for an automatic bank teller unit. (Mutual Federal Savings Bank)

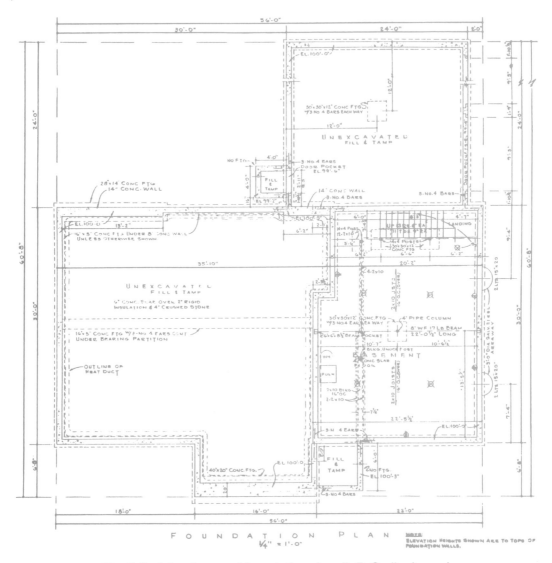

FOUNDATION PLAN
¼'' = 1'-0''

Fig. 8-6. A footings and foundation plan. (L.F. Garlinghouse)

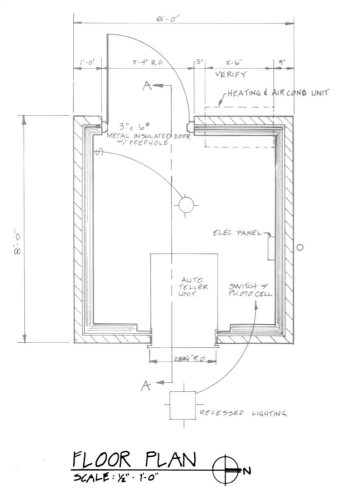

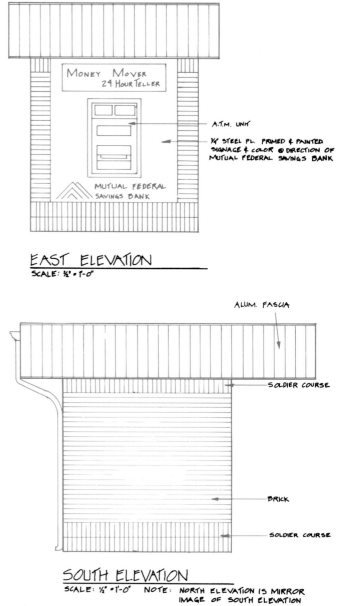

Fig. 8-7. A floor plan shows room placement and size. (Mutual Federal Savings Bank)

Fig. 8-8. Elevation drawings show what the building looks like from the outside. (Mutual Federal Savings Bank)

The *floor plan* is a fully dimensioned room layout of each floor, Fig. 8-7. All openings such as doors, windows, and walls are shown. The mechanical systems may be on the floor plan, too. They include the electrical, plumbing, and heating/cooling/ventilating systems. A floor plan is made for each level. The basement is drawn if it is finished. (A finished basement has floor, wall, and ceiling surfaces added to it.) Unfinished basements do not require a separate floor plan. The footing and foundation plan is used.

Elevation drawings are drawn to show the outside of the structure. These drawings appear as though you are looking straight at the structure. See Fig. 8-8. An idea of how the complete structure looks is shown in an elevation. They show the length and height of a house. All sides that are different are shown. Elevations of buildings show all grade, floor, and ceiling levels.

Sections and details show how the walls, ceiling, and roof are built. See Fig. 8-9. Detail drawings show fireplaces, stairways, and finish millwork.

WORKING DRAWINGS FOR LARGE PROJECTS

More than one set of working drawings is needed for large projects. Such projects will have a set of civil, site, architectural and engineering, mechanical and electrical, and structural drawings.

Civil drawings describe the site before building begins. Earthmoving and site preparation are described in Fig. 8-10. Civil drawings show changes in the course of a stream or road.

Site plans show drainage of excess water from the site and include building placement and grading

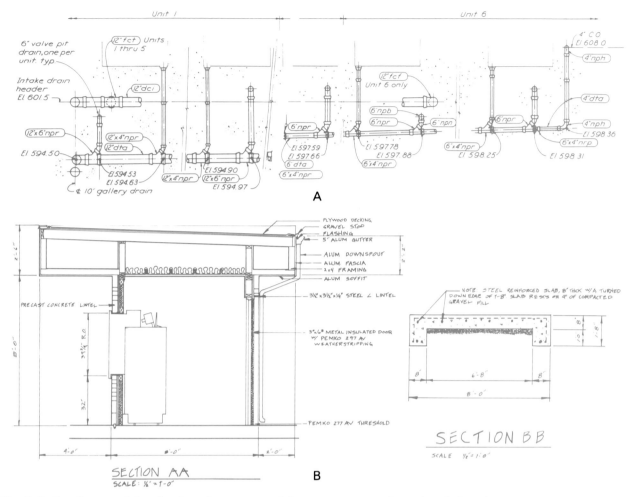

Fig. 8-9. Sections and details show how parts of the structure are shaped and placed.
A–Detail of plumbing system. B–Section AA shows a cut through automatic
teller building. Section BB shows side view of a cut through footing.
(Mutual Federal Savings Bank)

elevations. The final contour and landscape plan are in this set. See Fig. 8-11.

Architectural and engineering drawings are made to show the physical form of the project, Fig. 8-12. Many detailed drawings are needed. *Architectural drawings* are used to check the structure itself, Fig. 8-12, and add detail to master plans. *Engineering drawings* are used for utilities, roads, pipelines, and other long, narrow projects.

Mechanical and electrical drawings are used by electricians, plumbers, and heating/cooling contractors, Fig. 8-13. Mechanical drawings are important for large projects that have transportation systems in the project. *Mechanical drawings* describe the heating, ventilating, and air conditioning (HVAC) and plumbing systems. Elevators and moving stairs are common to public buildings with more than two floors. Mechanical drawings are used to describe these transportation systems. *Electrical drawings* show the kind and size of electrical

circuits and where to run wires. The type, size, and location of each electrical fixture is shown.

Structural drawings, Fig. 8-14, describe the main support system for a structure. Many detailed drawings are used. They show the size and shape of parts, and how parts are joined with other parts.

SPECIFICATIONS

Specifications are details about the project that are easier to put into words than into drawings. If drawings and specifications do not agree, specifications are followed.

Specifications are first used by contractors to estimate and bid the project. Specifications become a part of the contract when the contract is signed.

Specifiers write the specifications. They are guided by the project designers. Specifiers need to be fast, clear, and thorough. Methods have been

A

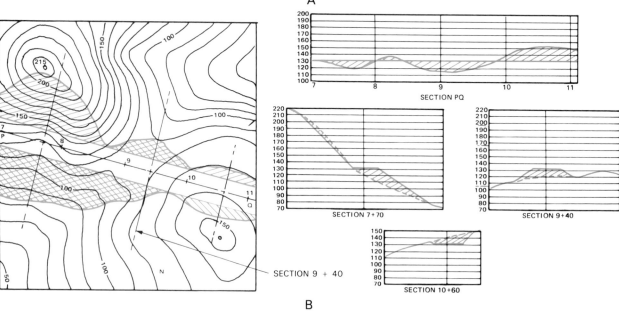

SECTION PQ

SECTION 7+70

SECTION 9+40

SECTION 9 + 40

SECTION 10+60

B

Fig. 8-10. A–This machine operator is moving earth. B–The kind of earth and where to put it are shown on the civil drawings. (Caterpillar Inc.)

Fig. 8-11. The workers are building a roadbed. A site plan shows the workers where to remove and dump the soil. (Caterpillar Inc.)

Fig. 8-12. This grader operator is using laser equipment to check the elevations while he works. (Spectra-Physics)

Fig. 8-13. Mechanical and electrical drawings are used by this electrician. They show him how to wire this light. (Leviton Manufacturing Co., Inc.)

developed to reach these goals. Specifiers can use copies of earlier work or use standard forms. Computers save a lot of time. Skilled specifiers are able to do the work right on the computer screen.

Organizing Specifications

The Construction Specifications Institute (CSI) describes how to organize and write specifications.

A list of titles and numbers are used to organize specifications. The system has 16 *divisions.* Refer back to the specification section of Fig. 8-15.

Division 1 describes general procedures. See Fig. 8-16. The other 15 divisions relate to construction standards for special trades. Fig. 8-17 is in reference to Division 3-Concrete. Refer back to the specification section in Fig. 8-15 for the names of the other divisions.

A

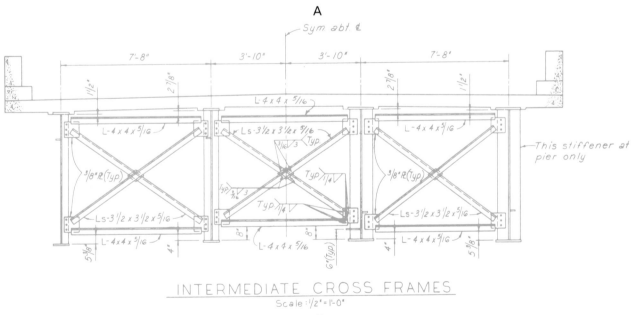

INTERMEDIATE CROSS FRAMES
Scale : 1/2" = 1'-0"

B

Fig. 8-14. A–Structural drawings describe these beams. Details show how they are joined.
B–Structural drawing of a bridge cross frame. (Corps of Engineers, Kansas City District)

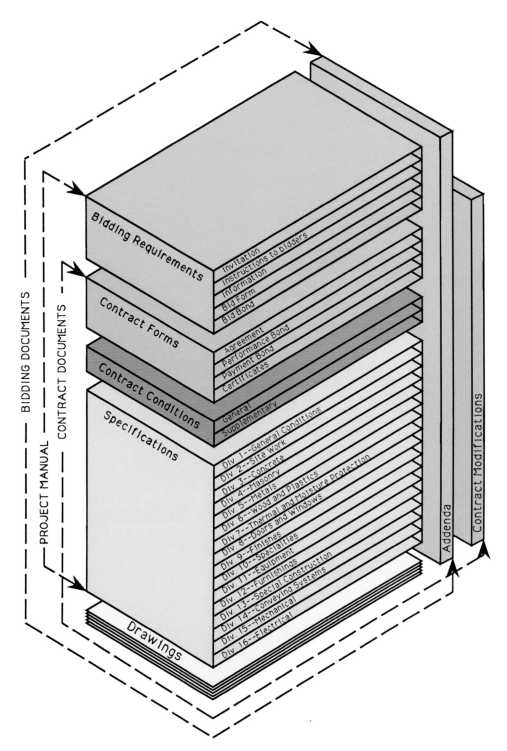

Fig. 8-15. Construction documents are packaged in different ways. Bidding documents are different than contract documents. The 16 divisions of the specifications make up one part. (Construction Specifications Institute)

Writing Specifications

Specifiers must decide on the method, scope, and the level of restrictiveness to use. The methods determine how the quality will be stated. Will it be based on performance of the product, a standard, or a brand name of a product?

The scope is used to describe the breadth of the system. A broad scope system may be the entire ventilating system for a highway tunnel. A narrow scope would refer to the intake and exhaust grates for the system.

The level of restrictiveness can limit or broaden the number of bidders. Flooring materials for a

Fig. 8-16. The specifications say this contractor cannot stop traffic while working. (CMI)

gym may be restricted to wood only. This would exclude all other kinds for flooring materials.

Matching Drawings and Specifications

Drafters and specifiers try hard to reduce errors. They work to reduce duplication. Duplication means to state something in more than one place. Details may differ if they are stated twice. Identical terms are used in both drawings and specifications.

Ways are found to resolve conflicts. Specifications generally are used when they differ from the drawings. An A/E may also be asked to put a ruling in writing.

Coordination guides are stated and followed to prevent errors. They list all standard terms and symbols that are to be used. The procedures used to check work and to settle conflicts are described.

Addenda

Many people study the documents during the bidding process. This often reveals oversights. These things need to be explained, corrected, or added. The addenda section describes all of the changes that are made in the documents.

Fig. 8-17. This concrete is special. Specifications tell how to make it and how strong it must be. (CMI)

Construction Standards

Construction standards describe the materials and equipment used in the project and the methods used to do the work at each stage of construction.

Materials

Quality of the materials used for the project is described. When the contractor selects materials, they are to be new and of the right grade for the purpose. The contractor must be able to prove the quality of materials used. The concrete in Fig. 8-17 must meet specifications. Samples must be taken and tested. In other cases, a stamped grade number for material is specified on drawings.

Equipment

Equipment used in the project must work to the level stated in the specifications. It must be approved by the engineer. Catalogs and data sheets are used for this purpose. The equipment shown in Fig. 8-18 was approved by the engineer.

Equipment installed as part of a factory must also be approved. The drawing will often specify a brand name for this equipment. The brand is known to have high quality.

The owner may accept more than one material or machine for the project. For example, in a sewer project two or more kinds of pipe may be used. Ductile iron, or precast concrete, clay, or plastic pipes may be chosen. The specifications describe the pipes and how to install them.

Methods

Contractors may use more than one method of construction in addition to choices in materials and equipment. There is more than one way to place sewer pipes under railroads without moving the tracks. You can tunnel, bore, or jack (push) the pipe under. The contractor can decide how to do it. The specifications describe each method and what the engineer will accept.

SUMMARY

Working drawings and specifications describe how to build a project. They are used by contractors

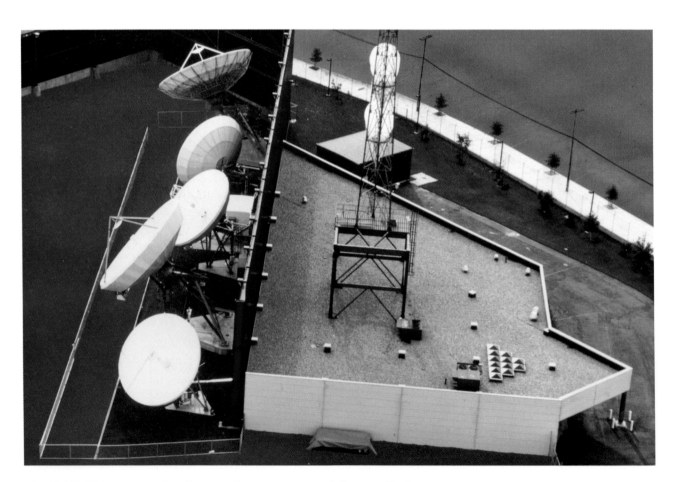

Fig. 8-18. This communications equipment was carefully specified. It must perform to a standard. (Hughes Communication, Inc.)

to estimate and bid a project. They become a part of the contract when it is signed.

Drawings are put into sets. The set describes how to prepare the site, the form of the project, the mechanical systems, and the structural framework. Shop drawings are made for the subcontractor.

Specifications are written out. They describe the bidding requirements, contract limits, and construction standards.

KEY WORDS

All the following words have been used in this chapter. Do you know their meaning?

Architectural drawings
Civil drawings
Construction standards
Divisions
Electrical drawings
Elevation drawings
Engineering drawings
Floor plan
Footings and foundation plan
Landscape plan
Mechanical drawings
Rebar
Sections and details
Shop drawings
Site plan
Specifiers
Structural drawings

TEST YOUR KNOWLEDGE

Write your answers on a separate sheet of paper. Do not write in this book.

1. An offer to build a project at a stated price is a _____.

2. A contractor may make shop drawings for a _____.

3. Which of the following is not shown in a site plan?
   a. Final shape of the site.
   b. Section view of foundation.
   c. Placement of structures.
   d. Boundary lines.
   e. Easements.

4. More than one set of working drawings are needed for _____ projects.

5. A _____ drawing describes the site before building begins.

6. What are specifications?

7. A _____ in the contract limits states what the contractor pays if a deadline is missed.
   a. bidding requirement
   b. default
   c. change order
   d. penalty clause

8. Materials for the project are to be _____ and of the right grade.

9. Catalogs and _____ sheets help the engineer approve equipment used to build a project.

10. The contractor must be able to _____ the quality of materials used.

ACTIVITIES

1. Visit an architect or engineer. Ask to see a set of drawings for a small project.

2. Visit the waterworks or waste water treatment plant in your community. Ask to see a set of drawings and specifications. Read through the specifications. Can you find the bidding requirements, contract limits, and construction standards?

3. Go to the building supply store. Get a set of drawings for a storage shed. Write down what you understand about sizes of its parts and materials used.

Section 4
MANAGING CONSTRUCTION ACTIVITIES

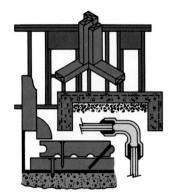

Chapter 9
CONTRACTING

After studying this chapter, you will be able to:
☐ Compare a general contractor, subcontractor, and supplier.
☐ Choose a price-setting method for a contract.
☐ Give advantages of verbal, negotiated, and competitive bid contracts.
☐ Tell what the bid opening meeting involves.
☐ Describe the importance of an estimator in creating a contract.
☐ List some things that raise labor costs and material costs.
☐ Define the terms overhead and profit.

Projects are built by owners, architects/engineers, or contractors. Owners build their own projects when structures are small. Firms with a lot of work may have their own crews. Most work done by owners relates to repairing (fixing), altering (changing the use), or modifying (changing the size) of their structures.

The A/E may agree to build the project. A designer/builder is called a *constructor,* Fig. 9-1. A constructor is a firm that designs and builds a project for a fee.

Constructors are able to organize the total project. The owner saves the time and expense of getting a contractor. But, the building costs may be higher because builders do not compete for the work. Some owners feel it is better for the designers and builders to work as a team on very large projects. Fig. 9-2 provides an example of one such large project.

Most projects are built by *contractors* who compete for the projects. This method has worked well through the years. The owner is able to get the project built right at the least cost. This method makes builders produce more at less cost in order to earn a profit.

KINDS OF CONTRACTORS

There are three kinds of contractors: general contractors, subcontractors, and suppliers. *General contractors* get the overall contract, Fig. 9-3. The overall contract includes all work done. General contractors manage construction from start to finish, Fig. 9-4.

An owner can be a general contractor. A family that builds its own house is like a general contractor.

The family may subcontract building the basement and the plumbing. The rest they do themselves. In this way, the family saves the money for the labor and the profit that would go to a general

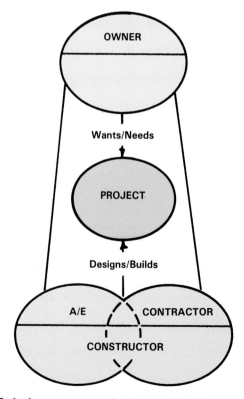

Fig. 9-1. A constructor designs and builds a project.

Fig. 9-2. This new port was built by a constructor. The volume and kinds of cargo to be handled were given to the constructor's engineers. Then, the firm designed and built the port. (Raymond International, Inc.)

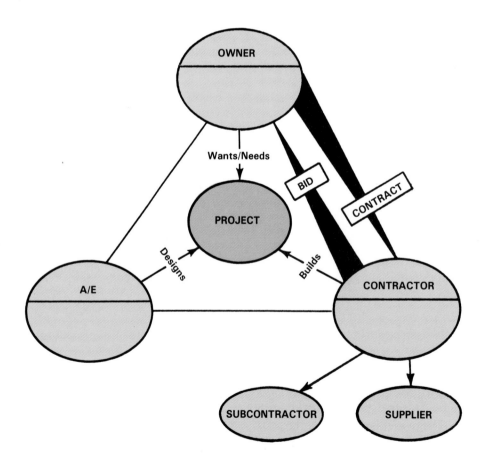

Fig. 9-3. The general contractor decides what the subcontractors and suppliers do.

Fig. 9-4. The general contractor for this highway built the roadbed and paved the road. A subcontractor built the bridges. (Department of Transportation, Federal Highway Administration)

Fig. 9-5. This subcontractor drills water wells. The company drills a hole, puts a casing into the hole, and installs a pump. (Ingersoll-Rand Company)

contractor. The family also must accept the risk of the work. Other owners can do the same thing. It requires skilled people and a desire to accept the risks.

Subcontractors do only one kind of work, Fig. 9-5. A subcontractor may erect steel frames. Another builds concrete forms, places concrete, and removes forms. In putting up a building, many special crafts are needed. There are fewer special crafts needed to build pipelines and roadways.

Suppliers provide materials and equipment. A contract states the price and delivery schedules. Fig. 9-6 shows a supplier making a delivery. A builder of a skyscraper has many suppliers. A highway builder does not have as many.

WHAT IS A CONTRACT?

A *contract* is a detailed agreement between two parties (people or firms). The contract has *terms.* One of the terms describes the work to be done. Working drawings and construction standards describe the work.

The price is one of the terms of a contract. The *price* is the amount of money paid to get the work done, and it is set before work begins.

The price for a project is set in four ways: the fixed price method, the cost plus fixed fee contract, the cost plus percentage of cost, and the incentive contract.

Fig. 9-6. This concrete supplier must supply the right amount of the right mix when and where the contractor wants it. (DYK Prestressed Tanks, Inc.)

In the *fixed price method,* the builder agrees to do all of the work for a fixed price. One amount of money covers the cost of materials, labor, equipment, overhead (office expense), and profit. The money can be paid as a single lump sum when the project is done.

Larger fixed price projects may have a unit payment plan. The *unit payment plan* states that the contractor is to be paid a set amount when certain stages of the project are complete.

A partial payment may occur when the site has been cleared. Another may be made when the foundations are complete.

A payment schedule can also be based on *partial payment estimates.* At stated times, the contractor can receive payment for the materials and labor that have been used on the project. Part of the payment may be withheld until the project is done. The amount held back and time of payment are both stated in the contract.

The *cost plus a fixed fee (CPFF)* contract lowers the risk for the builder. Materials, labor, and equipment make up the cost. The fixed fee is to cover overhead and profit.

The fixed fee is written into the contract. The costs are added up while building. There is little reason for the builder to control building costs.

Large industrial projects are carried out using the CPFF contract. The builders are chosen because they are able to keep costs down.

The *cost plus percentage of cost (CPPC)* is rare. This kind of contract has the owner pay for all of the materials, labor, and equipment. The contractor's profit and overhead is paid for with a percentage of the costs.

Like the CPFF contracts, the owner following CPPC pays for all building costs. The contractor's overhead and profit is based on a percentage. This contract method is seldom used because slow, wasteful work habits result in more profit. The higher the costs for materials, equipment, and labor, the more the contractor earns. In many states this form of contract is against the law. This kind of contract lets building begin before the project is fully designed.

The last kind of contract is the incentive contract. With the *incentive contract,* the owner and contractor agree on a target estimate of the cost and time to complete the project. The contractor and owner share in the savings if the real cost is less than the estimate. If the costs go over the estimate, some of the extra cost comes from the contractor's fees.

Getting the project done early can result in a bonus for the contractor. If it is late, a penalty is paid.

Other terms describe the payment schedule, warranties, construction schedules, and completion dates. Many details are in contracts. See a lawyer before signing a contract.

KINDS OF CONTRACTS

Contracts may be verbal, negotiated, or competitive. *Verbal contracts* are made with words and maybe a handshake. No papers are signed. A verbal contract is often used for small building projects.

It is best to draw up a written contract. The *written contract* is signed by each party. If a contract is written right, it will hold up in a court of law. Verbal contracts may not.

Negotiated Contracts

Negotiated contracts result when the owner and contractor talk about the work and price. This kind of contract is common to private owners. An owner may talk with more than one contractor to find the lowest price. Money and time are saved by negotiating the contract. Time used for the bidding process is saved. This time savings allows the owner to get started building sooner. Money can be saved if shortcuts are discussed.

Competitive Bidding

Two or more contractors can offer to build the project. Each contractor tries to get the job. They are said to compete for the job. This is called *competitive bidding.*

Private owners may feel it is best to use competitive bidding to find a builder. Since the money they spend is their own, they will try to get the most value from it.

Private owners may select two or more contractors they feel can build their project. These contractors are asked to bid on the project. The process varies greatly. The only rules private owners have to follow are rules of fair play.

Public owners are bound by local, state, and federal laws. The common process has seven steps. See Fig. 9-7.

Advertise for bids

On public projects, anyone can bid on a project. Since all contractors pay taxes, they should all have a chance to get a contract for a public project. An

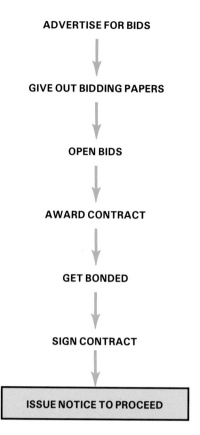

ADVERTISE FOR BIDS

↓

GIVE OUT BIDDING PAPERS

↓

OPEN BIDS

↓

AWARD CONTRACT

↓

GET BONDED

↓

SIGN CONTRACT

↓

ISSUE NOTICE TO PROCEED

Fig. 9-7. Competitive bidding for a public project follows a procedure.

effort is made to let all contractors know about the project. Notice is given through newspapers, magazines, and contractor associations. Special notices are sent to qualified builders. A *qualified builder* is one who is capable of building the project.

The *notice* describes the project. It states the type of work, location, and time and place of bid opening. It also states where and how to get the documents needed for bidding.

Private owners may select qualified contractors at this point in the process. Only those who are able to build the project are asked to bid.

Give out bidding documents

Bidding documents are the link between the A/E and the builder. See Fig. 8-15. Contractors first use bidding documents to guide them in submitting bids. A set of bidding documents consists of:
• Bidding requirements.
• Contract forms.
• Contract conditions.
• Specifications.
• Drawings.
• Addenda.

Bidding requirements. *Bidding requirements* describe how contractors are to submit their bids. A

formal *invitation to bidders* comes first. It describes the nature of the project. It is followed by specific *instructions to bidders.* All of the *bid forms* a prospective bidder will need are included.

The size and kind of a bid bond is described. A *bid bond* is a form of insurance. The bidder has to write a check for a set amount. The check is sent in with the bid. The owner does not cash the check. If the bidder is awarded the contract, but is not able to provide the service, the owner can then cash the check.

Contract forms. A copy of the *agreement* (contract) between the owner and contractor is enclosed. Many kinds of standard agreement forms are published.

The nature and size of bonds the contractor must have are described. These *bonds* protect the owner if the contractor defaults (breaks) the contract. They assure the owner that the project will be finished and that all bills will be paid.

Copies of all *certificates* used to show that the work is started, approved, and finished must be attached.

Contract conditions. The rights and responsibilities of the owner, designer, and contractor are stated in the contract conditions. These written conditions have been revised and improved over a period of many years. *General conditions* apply to nearly all construction projects. *Supplementary conditions* vary from project to project.

Working drawings, specifications, and addenda. Working drawings, specifications, and addenda tell the builder how the A/Es want the projects built. These documents were described in Chapter 8.

Open bids

The owner, the A/E (representing the owner), the contractor, and legal councils of both parties are present when they open the bids. Any parties may have other people represent them when officials open the bids. Contractors at a bid opening are shown in Fig. 9-8.

Anyone can attend a bid opening for public projects. They can also look at each bid. The owners of a public project include the person in the top city office (most often the mayor), the city council, and the people who manage the new project when it is built.

When the officials open the bid, the rules state the following:
• Bids are to be in sealed envelopes.
• Bids cannot be changed after the time of opening.

Fig. 9-8. Contractors attend bid openings like this one. All firms are prequalified. Contracts are given to the lowest bidder. (Colorado Dept. of Highways)

- Bids for public projects are read aloud.
- The *lowest responsible bidder* is given the contract. (A *responsible bidder* is one who is able to build the project and has made a correct and complete bid.)

When the officials open the bids, they check each bid to see if it is in order. Each bid must contain the bid bond. A bidding form must be complete and correct.

Officials will not accept any changes in the drawings or specifications by the bidder. For example, latex wall paint cannot be used instead of oil base wall paint.

The math is checked to see if it is correct. The bids are given to the A/E. A recommendation is made by the A/E to the owner.

Award contract

The lowest responsible bidder is sent a *notice of award.* The award is not a contract. It becomes final when the owner is able to get the money to build and the contractor gets bonded.

Bonding

The contractor needs to get a performance bond and a payment bond. Bonds are a kind of insurance. A *performance bond* insures that the project will be built if the contractor defaults on the contract. If the contractor has a *payment bond,* all workers and suppliers will be paid if the contractor goes bankrupt (goes into debt).

Sign contract

The owner and contractor sign a contract. The contract includes the working drawings, specifications, bonds, payment schedule, and all other terms.

Issue notice to proceed

The law firm sends the contractor a *notice to proceed.* This paper lets the contractor begin building. It also sets the starting and ending days for the project.

ESTIMATING

Early in the planning, owners and the A/E make rough cost estimates. An *estimate* is the amount of money a contractor thinks it will take to build a project. The contractor arrives at the estimate by adding up all of the costs for materials, labor, equipment, overhead (office expense), and profit.

An *estimator* is a key person in a construction firm. See Fig. 9-9. The success or failure of a firm depends on the skill of the estimator. If the estimate is too high, the contractor will not get enough contracts. If the estimate is too low, there will be too little profit.

Fig. 9-9. Being able to read drawings and make correct ''take-offs'' and figure the costs is a must for an estimator.
(International Conference of Building Officials)

Fig. 9-10. Estimators can make or break a contractor. There are many chances for error on this project.
(Raymond International, Inc.)

The goal is to make a fairly good guess of the cost within the time given. Whether the project is a home or a dam, the process is the same. First, look at the working drawings. Second, break down all work into a sequence of units. A unit is a part of the project. Third, compute the cost of each unit of work.

Good estimators:
- *Know* the construction business.
- *Read* working drawings.
- *Make* correct take-offs (count number of pieces, feet, etc.).
- *Plan* a good building sequence.
- *Keep* unit cost records of past projects.

Making Estimates

There are several areas of cost that the estimator must judge when making an estimate. These important areas are: materials, labor, equipment, and profit.

Materials

You estimate materials by counting the number of each piece that is needed in the project. Counting the number of each size of steel for the offshore coal terminal in Fig. 9-10 takes care.

Concrete is bought by volume. The estimator will need to compute the number of cubic yards (a space 3 x 3 x 3 ft.) of concrete needed in the project. Fig. 9-11B shows you how to figure the volume of a

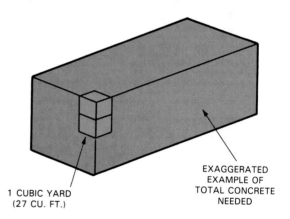

EXAGGERATED EXAMPLE OF TOTAL CONCRETE NEEDED

1 CUBIC YARD (27 CU. FT.)

$$\frac{\text{Thickness (T) ft.} \times \text{Width (W) ft.} \times \text{Length (L) ft.}}{27 \text{ feet}^3} = \text{cubic yards (cu. yd.)}$$

$$\frac{15 \text{ ft.} \times 20 \text{ ft.} \times 60 \text{ ft.}}{27 \text{ ft.}^3} = \frac{18,000 \text{ ft.}^3}{27 \text{ ft.}^3} = 667 \text{ cu. yd.}$$

A

$$\frac{1/3 \text{ ft.} \times 3 \text{ ft.} \times 54 \text{ ft.}}{27 \text{ ft.}^3} = \frac{54 \text{ ft.}^3}{27 \text{ ft.}^3} = 2 \text{ cu. yd.}$$

B

Fig. 9-11. The formula used to figure the volume of concrete. A–An exaggerated example. B–A common sidewalk problem.

sidewalk that is 4 in. thick, 3 ft. wide, and 54 ft. long. To find the cost for the concrete, take the number of units (cubic yards) times the cost per unit.

The formula used to figure the cost of concrete when the volume is known is the following:

Number of Cubic Yards (cu. yd.) x Cost/cu. yd.
= Cost of Concrete

2 cu. yd. x $80.00/cu.yd. = $160.00

The concrete for the sidewalk will cost $160 if the concrete is $80 per yard.

Labor

Cost of labor is found by estimating the number of hours it takes to do a job. This number is taken times the rate per hour paid to each worker.

In the sidewalk example, the estimator adds up the time it takes for each part of the job. Time is used to:

• Remove and dispose of the old sidewalk.
• Prepare the base, build forms, and place wire fabric.
• Place, finish, and cure the new concrete.
• Remove forms and clean up the site.

Time estimates can be taken from tables. *Labor productivity tables* are available for almost all building tasks. If careful records are kept on projects, an estimator can use the worker's own rates.

Using the estimator's own records is best because the rate of each job causes the worker's rate to differ from the average.

Equipment

The cost per hour to run equipment is used to figure equipment costs. See Fig. 9-12. If the equipment is rented, the contractor pays for rent, fuel, repairs, and any steps needed for moving it.

The total cost is divided by the hours worked. In this way, one gets an estimate of the *cost per hour*. If the total costs are divided by units of work done, the answer would be the *cost per unit of work*. The cost of digging the trench in Fig. 9-13 was figured that way.

Owners of equipment must figure the cost of depreciation (loss of value), interest, taxes, insurance, storage, fuel, and repairs. *Depreciation* is the loss in resale value of the equipment as it gets older. Interest is the money paid to a lender for the money used to purchase the equipment.

Overhead

Overhead is the cost of running a firm. There are two kinds of overhead. The first is home office overhead costs. See Fig. 9-14. *Home office overhead* includes the payroll for those who manage and

Fig. 9-12. The cost of moving earth is hard to estimate. The volume and kind of soil along with the length of haul must be figured. (Caterpillar Inc.)

Fig. 9-13. The cost of digging this trench is estimated in dollars per foot. The estimator adjusts the estimate for the kind of soil and depth of trench. (Barber-Greene Co.)

Fig. 9-14. These people work in the contractor's central office. Their salaries are part of the overhead. (Honeywell, Inc.)

work in the office. Office rent, telephone service, and utilities add to overhead expense as does advertising and travel.

Field office overhead costs are added to each job, Fig. 9-15. *Field office overhead* includes the payroll for everyone except the workers. Surveys, office space, testing, site preparation, insurance, storage, building permits, and payments for bonds all add to field office overhead costs.

Fig. 9-15. Getting power to the site and setting up the field office takes money. These costs and others are field office overhead. (DYK Prestressed Tanks, Inc.)

Profit

If a contractor is to stay in business, he or she needs to make a profit. The difference between what the contractor gets for building the project (the receipts) and the builder's costs, is *profit.*

Contractors look at many factors when they decide on the markup on a project. *Markup* is an estimate used to plan the level of profit.

Some questions to ask before setting the markup are listed below.

• Can I control the risk in the job?
• Do I want to enter this new market?
• Does the job use key people who are available?
• Are prices stable?
• Do I get paid as I go?
• Can I group my purchases on other projects and get quantity discounts?
• Do I need the work?

How these questions affect the markup is shown in Fig. 9-16.

Fig. 9-16. YES answers mean that a smaller markup (profit) needs to be added. NO answers mean larger profits are needed. Can you explain why?

An estimator computes the cost of material, equipment, subcontractor's quotes, and overhead. Top management decides on the amount of markup and prepares the bid. Fig. 9-17 shows this concept. Computing building costs is a matter of following procedures. The profit decision is a matter of strategy. A strategy is a plan to reach a goal. Choosing a strategy is the job of the firm's top leaders.

SUMMARY

Contracting is complex. It involves many kinds of contracts. At each step, you must attend to details. A contractor must follow a set process. Large sums of money are at stake and care must be used to prepare the bids. Accurate estimating is vital to become a successful contractor.

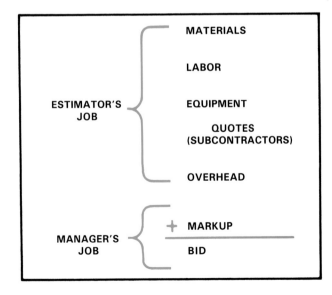

Fig. 9-17. The work of the estimator is routine. Top management decides on the amount of markup and prepares the bid.

KEY WORDS

All the following words have been used in this chapter. Do you know their meaning?

Agreement
Bid bond
Bid form
Bidding documents
Bidding requirements
Bond
Certificate
Competitive bidding
Constructor
Contract
Contractor
Cost per hour
Cost per unit of work
Cost plus a fixed fee (CPFF)
Cost plus a percentage of cost (CPPC)
Depreciation
Estimate
Estimator
Field office overhead
Fixed price method
General condition
General contractor
Home office overhead
Incentive contract
Instructions to bidders
Invitation to bidders
Labor productivity tables
Lowest responsible bidder
Markup
Negotiated contract
Notice
Notice of award
Notice to proceed
Overhead
Partial payment estimate
Payment bond
Performance bond
Price
Profit
Qualified bidder
Subcontractor
Supplementary condition
Supplier
Terms
Unit payment plan
Verbal contract
Written contract

TEST YOUR KNOWLEDGE

Write your answers on a separate sheet of paper. Do not write in this book.

1. A designer/builder is called a _____.
2. List the three types of contractors.
3. A _____ plan states that the contractor is to be paid a set amount when certain stages of the project are complete.
   a. partial payment
   b. CPFF
   c. incentive
   d. unit payment
   e. CPPC
4. A (written/verbal) contract will usually hold up in a court of law.
5. What is competitive bidding?
6. To inform many contractors of a future job, a public owner will _____ for bids.
7. A bid bond assures that the bidder accepts the contract if an offer is made. True or false?
8. The time given to prepare bids ranges from _____ to _____ days.
9. At a public bid opening, bids cannot be _____ after the time of opening.
10. When the lowest bidder receives a notice of award, the award (is/is not) a contract.
11. What do you call the amount of money a contractor thinks it will take to build a project?
12. List some steps needed for good estimating.
13. If an estimate is too low, profits will be too (high/low).
14. Labor _____ tables are used to estimate labor costs.
15. If the total labor cost is divided by the hours worked, the result is the _____ _____ _____.

Matching questions: On a separate sheet of paper, match the definition in the left-hand column with the correct term in the right-hand column.

16. _____ Costs of surveys, site preparation, insurance, and building permits.
17. _____ A plan to reach a prearranged goal.
18. _____ Includes office payroll, telephone.
19. _____ An estimate used to plan profit level.
20. _____ Receipts minus costs.

a. Home office overhead costs.
b. Field office overhead costs.
c. Profit.
d. Strategy.
e. Markup.

ACTIVITIES

1. Much of what builders do is controlled by verbal (spoken) and written contracts. Think of examples.
2. Do you or the people you live with have contacts with employees or unions? What are the terms of the contracts?
3. Let us say you need a new garden shed. Plan it out and make some sketches. How much will the materials cost? How much time will it take to build it? Are there any overhead costs (gas for car to pick up supplies)? What will it cost to buy or rent equipment to build it? Add up all the costs. Make a written report for your teacher and an oral report to your class.

A   B

Types of contractors include mechanical contractors, electrical contractors, and others. A–Highway work is done by a general contractor who has bid for the project. (Austin Industries) B–Work for an electrical contractor involves commercial and residential wiring. (American Electric)

Chapter 10

MANAGING THE CONSTRUCTION PROJECT

After studying this chapter, you will be able to:
☐ Discuss terms like plan, schedule, direct, and control.
☐ Describe marketing, engineering, and the other departments.
☐ Discuss the bar chart and critical path methods.
☐ Start making a bar chart for a sample project.

How well people manage a project will affect the amount of profit or loss. Well-led projects are done on time. They are built properly at the least cost. The builder gets the job done, but does not exploit (overwork) the workers. Correct and safe methods are used to complete the project.

How and when the builder completes a project reflects on the builder's reputation. A reputation is what people think about him or her as a builder.

People will rely on the firm to do a good job on time.

WHAT IS CONSTRUCTION MANAGEMENT?

Those who manage construction do four things. They plan, schedule, direct, and control the work. *Planning* is dividing the work into parts. Let us pretend we are going to erect a basketball hoop and backboard on a pole. Each task is listed in Fig. 10-1A.

The tasks are not in order. Arranging them in the order they must be done is called *scheduling.* Fig. 10-1B shows the tasks as they might be scheduled.

As with most construction projects, there is more than one way to erect a basketball goal. The builder

TASKS	SCHEDULE
Lay out site	1. Lay out site
Buy goal and backboard	2. Buy goal and backboard
Buy pole, concrete, and steel	3. Buy lumber
Dig hole	4. Buy pole, concrete, and steel (all at one store)
Mix and place concrete	5. Dig hole
Set pole	6. Build form
Assemble goal and backboard	7. Set steel
Clean up site	8. Position pole
Set steel in hole	9. Mix and place concrete
Build form	10. Finish concrete
Strip form	11. Assemble goal and backboard
Buy lumber	12. Put goal and backboard on pole
Finish concrete	13. Strip form
Put goal and backboard on pole	14. Clean up the site

A   B

Fig. 10-1 . Planning and scheduling the basketball goal project.

uses the schedule that takes the least time to complete the project.

A second concern is to be sure there is work on the project for all workers. These two guides are based on two thoughts. First, the longer it takes to complete the project the more it costs. The second, idle workers are paid the same as busy workers.

When the schedule is ready, the person in charge of the project begins *directing* people to do the work. The worker is told what needs to be done. A worker may be told to use the spade to dig a hole that is 2 ft. long, 2 ft. wide, and 2 ft. deep. The location of the hole is marked. More instructions may be given, such as:

- The hole should have straight sides.
- All loose dirt must be taken out of the hole.

Time should be taken to explain why the hole should meet the standards given.

The next task of the leader is to control the work. *Controlling* means seeing that the work is done right and finished on time, Fig. 10-2. The purpose is to make sure the work is done right. Making sure that the work matches the drawings and the specifications is the second reason to check the work. A third is to give the worker feedback on how well the job was done. People will work better and harder when they know they are doing the work right and that the boss likes it.

Fig. 10-2. Checking the work of others is one part of the control function. Telling workers how they did is a second. (The Stanley Works)

HOW IS A CONSTRUCTION FIRM ORGANIZED?

A construction firm builds the project. People manage work in five activity areas. These areas are: engineering, production, marketing, financial affairs, and labor relations.

Activity Areas

Engineering provides the technical data for building work. Crews from engineering conduct soil and surface surveys of the site. They run tests on the soil and materials. Engineering designs the layout of the site so that work will progress as planned.

The *production* activity area is the source of the firm's income and profit. The service provided is the erecting of durable structures. All other activity areas support production. Most of the later chapters describe what is involved in production.

The *marketing* area promotes the firm's image and displays what the company does for others. People in this area prepare printed booklets that promote the firm. Marketing people try to convince project planners/owners that their firm is the best one to build the project.

Financial affairs manages the firm's money. People in the area watch the firm's cash flow. Cash flow is the money that is paid out and taken in. The people in financial affairs pay wages and purchase what is needed to build the project and run the firm.

Income comes from partial payments and contract closings. A partial payment is made as construction progresses. It is common on large projects. A *contract closing* occurs when a project is completed and the final payment is made. Accounting keeps records of the income and expense.

Labor relations settles problems for the firm's workers. This section works out contracts and sets wages. Workers in the labor relations section train and advance workers. Leaders from labor and the firm work as a team to promote better working conditions, safety, training, and more new projects. Both groups stand to gain in all of these efforts. Sometimes things do not work out, Fig. 10-3.

Levels of Authority

A construction firm has levels of authority. Firms are set up in many ways. The line and staff chart shown in Fig. 10-4 is used to describe large firms. One person may do more than one job in a

Fig. 10-3. When a contract or grievance is not worked out, a strike can occur.
(Muncie Newspapers, Inc.)

small firm. Larger firms may have more than one person working in each office at each level.

The owner of the firm can be one person (single proprietorship), two or more people (partnership), or many people (stockholders of a corporation). The stockholders elect a board of directors. This group of leaders decides on the goals of the firm.

The board selects a person to be president. This person has direct control over the firm. The goals of the firm direct the president's work.

One or more people carry out the work in each of the sections. The engineering section may consist only of the president in a small firm. In a larger firm there may be a staff of engineers. One group makes up shop drawings. Other people may do the estimating, site design, and other tasks.

It is the same way in the labor relations section. The leader on the site may hire the workers and

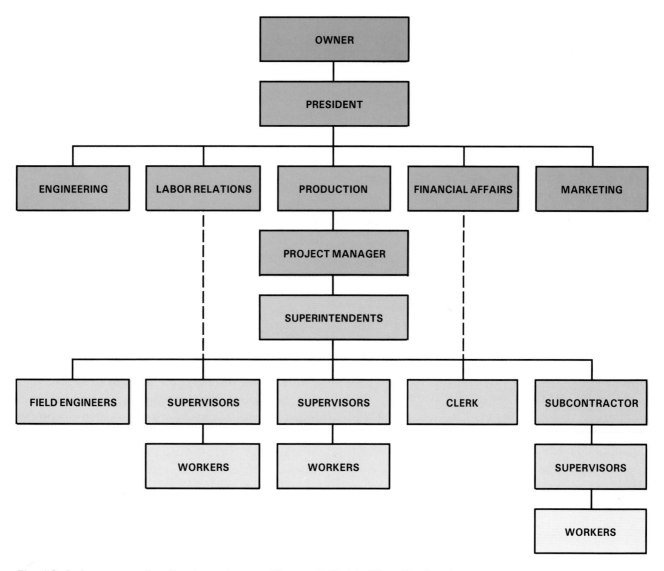

Fig. 10-4. A construction firm has a home office and a field office. The levels of authority are shown for all of the leaders and workers.

train them in a small firm. This task is done by the head of personnel in a large firm.

In the home office, work on the site is led by a *project manager.* On the site, building is led by the *superintendent.*

The project manager may be in charge of one large project or more than one smaller project. A project manager watches the progress of the project. The progress on the job is matched with the master schedule.

The project manager and the superintendent keep in close contact. Facts about costs, schedules, and equipment is given back and forth to keep work moving ahead.

The superintendent manages the field office. Large projects will have a field engineer. A *field engineer* is a person from the engineering section at the home office. Surveys and tests are done on the site by the field engineer's crew.

A *clerk* takes care of receiving reports, payrolls, time cards, and other paperwork in the field office. This work relates to the financial affairs section in the home office.

Each craft may have a supervisor. This means that plumbers, carpenters, electrical workers, and concrete workers each have a supervisor. The supervisor gets orders from the superintendent. Supervisors, in turn, instruct their workers.

Subcontractors receive orders from the superintendent. If they are a large subcontractor, they may have their own supervisor. The crafters receive orders from the supervisor.

CONSTRUCTION MANAGING

People manage from both the home office and the field office. Those who lead from the home office make up the master plan for the project. A *master plan* consists of a list of tasks and the schedule to get them done.

The superintendent works with the people in the home office on the master plan. A superintendent is the leader in the field office at the job site. This person plans, schedules, directs, and controls each of the major tasks on the master plan. Fig. 10-5 shows the results of a superintendent's work.

The architect/engineers inspect the project while it is being built. The A/Es work for the owner. The A/Es cannot change the methods used by the builder. They only make sure that the builder follows the drawings and specifications. Large projects will have a full-time inspector on the site.

Fig. 10-5. What do you think the superintendent did to make this job run smoothly? (Maine Department of Transportation)

The people in the home office use reports from the superintendent to check on the progress of the project. They ask:
- What work is done?
- What money is spent?
- What equipment is used?

The progress is compared with what was planned. They look for places where progress on the site and the planned progress are not the same.

The superintendent is told what work is running ahead of schedule and what is behind schedule. The work that is behind schedule becomes a priority. *Priority work* is done before other work. In this way the project can be brought back onto schedule.

A task that is behind schedule can be brought back onto schedule in one or more ways. Workers can be taken from a job that is ahead of schedule. More workers can be hired until they are caught up. Subcontractors may be hired to catch up on the work. Workers can be paid to work overtime. Workers on overtime are paid more per hour for working longer days or more days per week.

Leaders in the home office get materials, workers, and equipment. See Fig. 10-6. Materials used in the project are new and meet standards stated in the specifications. Materials used to build the projects (form bracing, scaffolds, and the like) may be used more than once. It is a *purchasing agent* who orders new materials or schedules used materials.

Fig. 10-6. Material for a sewerline is being delivered. It was ordered by the home office. (Caterpillar Inc.)

Fig. 10-7. Equipment on this job site costs many thousands of dollars. Those who manage the firm try to keep it in use.
(Michigan Department of Transportation)

Equipment is bought new, rented, or scheduled from other projects. Construction equipment is very costly. A major task is to keep it in use, Fig. 10-7. Idle equipment does not earn the builder any profit.

In construction, there are two kinds of workers. First, there are the workers who work for the general contractor. The second kind of worker is hired by the subcontractors. See Fig. 10-8. The head office leaders schedule the subcontractors.

The superintendent schedules the firm's workers. The challenge is to maintain the best work group size with the right craft skills. If a group is too large or too small, costs will increase. Costs will go up when workers are not doing what they are trained to do. If the workers are members of unions, they are not able to move from one craft to a second.

Lastly, home office leaders manage the *cash flow* from partial payments received, money the leaders borrow, or the firm's reserves (savings). Money going out is used to pay for materials, labor, equipment, and overhead.

HOW ARE CONSTRUCTION PROJECTS MANAGED?

There are three methods used to manage projects: experience, the bar chart method, and the critical path method. The first is not a planned method. The others are more formal.

Experience

The first method is called "experience." Those who have been builders for many years acquire a

Fig. 10-8. A subcontractor only builds the tank. The work was scheduled by the home office people. (DYK Prestressed Tanks, Inc.)

"sense" for planning. They know how things need to be done. They have a "feel" for scheduling. They "know" when work is lagging behind. This method is best used on small projects.

The techniques described in the next two methods are used when projects are large.

Bar Chart

A bar chart is easy to read. A *bar chart* shows the schedule for work to be done. It shows what is done and where there are delays. You can compare the planned schedule with the work on the site. The bar chart shows how much money has been spent on the project. It will, also, forecast income from the project. The above facts are needed by leaders so they can adjust the work arrangements.

Fig. 10-9 is a bar chart for the basketball goal project you learned about before. The bar chart was made this way:

1. List each task for the project.
2. Place the start times in order for each task.
3. Estimate the work time for each task.
4. Schedule the work.
5. Estimate the cost for each work task and write it in the value column. (Include labor, materials, and equipment.)

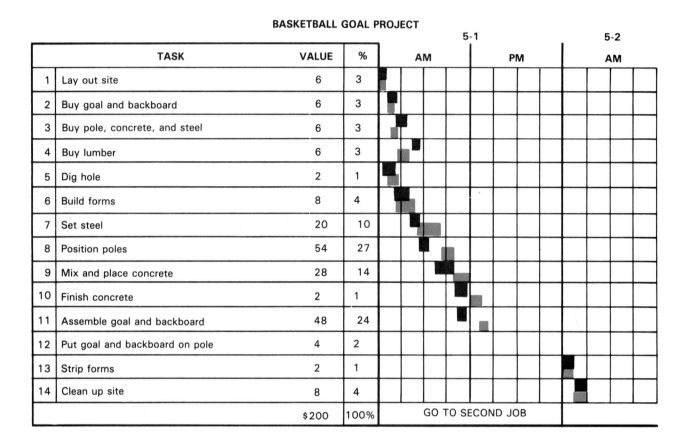

BASKETBALL GOAL PROJECT

	TASK	VALUE	%
1	Lay out site	6	3
2	Buy goal and backboard	6	3
3	Buy pole, concrete, and steel	6	3
4	Buy lumber	6	3
5	Dig hole	2	1
6	Build forms	8	4
7	Set steel	20	10
8	Position poles	54	27
9	Mix and place concrete	28	14
10	Finish concrete	2	1
11	Assemble goal and backboard	48	24
12	Put goal and backboard on pole	4	2
13	Strip forms	2	1
14	Clean up site	8	4
		$200	100%

GO TO SECOND JOB

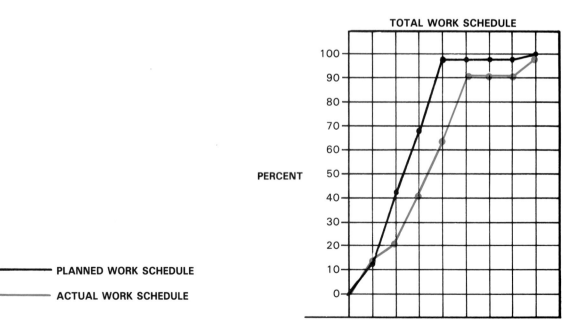

TOTAL WORK SCHEDULE

PERCENT

—————— PLANNED WORK SCHEDULE

—————— ACTUAL WORK SCHEDULE

Fig. 10-9. A bar chart of the basketball goal project.

6. Use the values from 5 above. Since each value is a part of the total cost, compute the percent of cost for each task. See Fig. 10-10.

7. Break down the percent of cost for each unit of time (day, week, month).

8. Develop a "Total Work Schedule." Refer back to Fig. 10-9.

9. Compute the percent of work finished for each task.

10. Add all of the percent quantities for the separate tasks to get a total completion quantity.

$$\% \text{ of costs} = \frac{\text{Task Cost}}{\text{Total Cost}} \times 100$$

$$\% \text{ of costs} = \frac{\$6}{\$200} \times 100$$

$$\% \text{ of costs} = 3$$

Fig. 10-10. How to compute the percent of cost for each task.

Do this for the planned and for the actual work completed.

11. Compute the actual work schedule at the end of each unit.

Critical Path Method (CPM)

The *critical path method (CPM)* shows the sequence of each task. Times are easy to add. The total time to build the project is easy to figure.

A CPM chart is an arrow chart. Fig. 10-11 is an arrow chart for the basketball goal project. Each arrow is a task. The circles with numbers in them are events. More than one arrow chart can be made. The best chart is one that gets the project done at the least cost.

A path is a series of events that are done in order. The path that takes the longest time to complete is the critical path (CP). This path controls the length of time needed to complete the project. If the job superintendent is able to shorten the CP, a second path may become critical.

The sequence that follows will help you make an arrow chart:

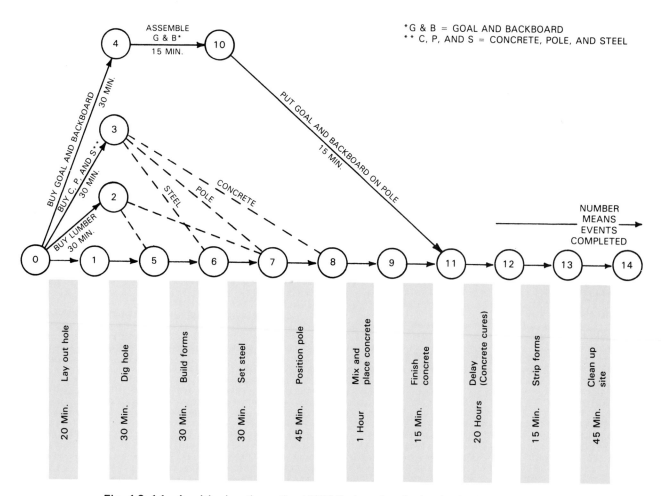

Fig. 10-11. A critical path method (CPM) chart for the basketball goal project.

1. List the events for the complete project.
2. List each task that leads to a complete event.
3. Write in the time needed to complete each task.
4. Make one or more arrow charts.
5. Draw the best arrow chart. (The one with the shortest CP.)

Computer systems make managing production easier, faster, and more precise. Computer software helps the planner organize the data or process. Programs remind the planner of steps that might be missed. They also calculate estimates and total times for operations. Printers and plotters produce printouts of tables, graphs, and charts. The planner uses these graphics to record data, check progress, and consider changes.

SUMMARY

Care in leading a project helps the firm earn more profit. Project leaders plan, schedule, direct, and control the project. These functions help to ensure a profit for the firm.

Experience, bar charts, and the critical path method are used by project leaders to plan work. These techniques help them get the project done on time. When work on a project goes smoothly, the owner, A/E, contractor, and workers all share in the money brought in.

KEY WORDS

All the following words have been used in this chapter. Do you know their meaning?

Bar chart
Cash flow
Clerk
Contract closing
Controlling
Critical path method (CPM)
Directing
Engineering
Field engineer
Financial affairs
Labor relations
Marketing
Master plan
Planning
Priority work
Production
Project manager
Purchasing agent
Scheduling
Superintendent

TEST YOUR KNOWLEDGE

Write your answers on a separate sheet of paper. Do not write in this book.

1. A manager must plan, _____, direct, and control the work.

Matching questions: On a separate sheet of paper, match the definition in the left-hand column with the correct term in the right-hand column.

2. _____ Provides the technical data for building work.
3. _____ Watches a firm's money and how it is spent.
4. _____ Works out contracts and wages for the firm's workers.
5. _____ Promotes the firm's image and its products.
6. _____ Concerned with erecting durable structures.

   a. Building activity area.
   b. Marketing.
   c. Financial affairs.
   d. Labor relations.
   e. Engineering.

7. The _____ agent orders new materials or schedules used materials.
8. The builder does not earn any profit from _____ equipment.
9. There are three methods used to manage projects: experience, the _____ _____ method, and the critical path method.
10. The path that takes the (shortest/longest) time is the critical path.

ACTIVITIES

1. Develop a bar chart for a new mailbox at your house.
2. Use the CPM to plan and schedule the building of a storage shed in your backyard. Make at least three arrow charts for comparing.
3. Use the library to find out how long some construction steps can take.
4. Use a computer program to plan the construction of a small project. Share your results with the class.

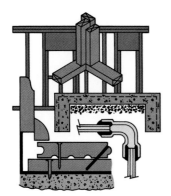

Chapter 11

GETTING WORKERS/CAREER CHOICES

After studying this chapter, you will be able to:
- ☐ Describe site work, building, finish work, and mechanical work.
- ☐ Discuss steps for hiring workers.
- ☐ State the functions of a labor union.
- ☐ Tell how to join a labor organization or union.
- ☐ List some careers available.

Working for a contractor differs from working for a manufacturer. Most construction work is short term. When the project is done, the worker has to find a new job. It is common for workers in a plant to be hired only once or a few times during their work life. It is just as common for construction workers to be hired five to eight times each year.

Construction workers must be able to adjust to new work rules and a new supervisor on each job. The duties that construction workers perform often change daily. Factory work has fewer changes.

Much of construction work is in the open or on a partly enclosed structure. Therefore, bad weather can stop the work. Fig. 11-1 makes this concept more clear.

KINDS OF WORKERS

Workers in construction do four kinds of work. They do site work, build the structure, do finish work on the structure, and do the mechanical work (plumbing, electrical, heating).

Site Work

Site work involves clearing away all objects or terrain not needed for the project. Survey crews use bench marks (existing survey markers) to locate structures, Fig. 11-2. Machine operators dig holes where footings or foundations will be placed. Sewers and

A

B

Fig. 11-1. A–Rain can bring work on this site to a halt. (Simpson Timber Co.)
B–The people in this warehouse may not even know it is raining. (Stran Buildings)

Fig. 11-2. Survey work on the site is a first step. Grade level and structure placement is done as soon as a site is cleared. (TOPCON)

A

B

other utilities are put under the ground. The work differs when the site is covered with water. Roadways have more site work to do than any other kind of construction.

Site work both starts and completes the project. When a project is nearly complete, the site is graded to get it ready for landscape work.

Building the Structure

The structure includes the substructure and the superstructure. The *substructure* is below the surface of the ground. Poured concrete footings and foundations are used for most houses. Pilings are used for heavy structures. The substructure for roadways may be packed soil or rock. Fig. 11-3 shows some kinds of substructures.

The *superstructure* for a building is attached to the foundation. It may be made of steel, concrete and steel, or wood. Frames give the superstructure shape and carry its load. Fig. 11-4 shows frames and assemblies in a structure. Paving, signs, and guardrails are highway superstructures.

Finishing the Structure

Finish work includes tasks which enclose and complete the project. Putting siding on the outside of a house is finish work. Covering walls, ceilings, and floors on the inside is also finish work. Laying sod and planting trees is doing finish work on the site. Fig. 11-5 shows two finish jobs.

C

Fig. 11-3. These pictures show how substructures are made. (A—Bob Dale, B—Hyster Co., C—Ingersoll Rand Co.)

Fig. 11-4. Can you find a substructure and a super-structure in this picture? (General Electric Co.)

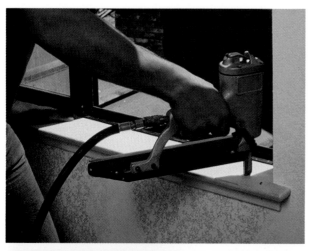

Fig. 11-5. These pictures show finish work. Explain why it is finish work.
(SENCO Products, Inc., California Redwood Assoc.)

Mechanical Work

Mechanical work includes putting in climate control equipment, plumbing systems, electrical power and lights, and elevators to transport people and goods.

Railroads have complex switching and signal systems. Highways have lighting and traffic signals. A pipeline is a mechanical system and little else. The same is true of plants that refine crude oil. Look at Fig. 11-6. Can you point out the mechanical systems?

Fig. 11-6. Which equipment moves air, water, or steam? (Honeywell, Inc.)

HOW ARE WORKERS HIRED?

Contractors find the people they need to build a project in several ways. The superintendent is most often a full-time employee of the firm. This person is assigned to a new project when the last one is complete.

When there is no work, superintendents will be kept on the payroll. They may be put on half pay until a new contract is signed. Sometimes they work on a job as a supervisor until they are assigned a new project. Others may work as crafters until the firm needs a superintendent.

A good supervisor may work as a crafter until the firm needs a supervisor. Good supervisors and skilled workers are of value to a firm. The firm tries to keep them as long as they can. Fig. 11-7 shows this concept.

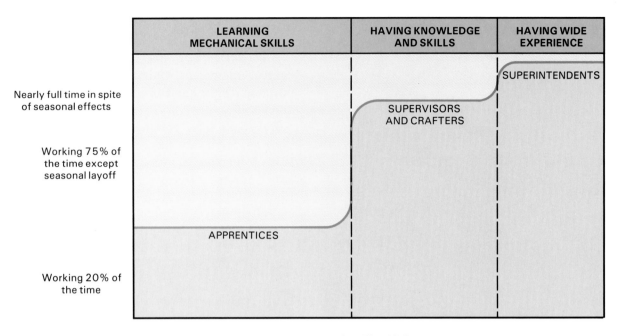

LEARNING MECHANICAL SKILLS	HAVING KNOWLEDGE AND SKILLS	HAVING WIDE EXPERIENCE

Nearly full time in spite of seasonal effects

SUPERINTENDENTS

SUPERVISORS AND CRAFTERS

Working 75% of the time except seasonal layoff

APPRENTICES

Working 20% of the time

Fig. 11-7. Most construction workers have temporary jobs. The higher the level the more permanent the jobs become.

Workers may live locally or be from out of town. These people may travel a long way to get the job.

The superintendent of a small project will hire the workers. In large firms, people from the personnel office do the hiring. They are a part of the labor relations section.

Hiring Process

People who hire others use a three-step hiring process. They recruit, then select, then induct workers.

Recruit

Recruiting, finding new employees, is done in four ways. A sign can be put up at the job site. People see the sign. They come in and apply for a job. This is called *walk-in application.*

The firm may advertise for help in local papers or trade journals. A person who wants the job applies in person or by letter.

A *job order* can be placed in an employment office. The office will advertise the work. They do much of the early screening. Only the most qualified workers are referred to the person doing the hiring.

The hiring process is often stated in union contracts. The union's business agent has a pool of skilled workers. When workers are needed, they are referred to the builder. This process helps the builder find skilled workers. It helps the worker, because they have a placement service.

Select

Next, the builder begins *selecting* the workers. They want the best workers. Good workers get more done and get it done right. This means more profit to the firm.

To make good choices, the person doing the hiring must have facts about the worker. They get the facts from the *job application.* They learn about past work, training, and skills of the person applying for the job.

Testing is used to find out what people know and what they can do. Paper and pencil tests provide facts about what they know. Work samples can show what they can do. Bricklayers may be asked questions about bricks and mortar on a paper and pencil test. In addition, they show what they can do by laying up a small section of a brick or block wall.

The person doing the hiring will interview the people with the most promise. An *interview* is a visit for discussion. More detail about the person's background is learned in the visit. People wanting the jobs have a chance to sell themselves. They try to convince the person of their desire to work and their value to the work force.

More facts can be found by checking the person's references. A *reference* is what other people say about the person applying for the job.

Induct

The personnel officer or the superintendent will then start *inducting* new workers. The first purpose is to help new workers feel they are a valued part of the team. They learn about the firm. Workers are told what the firm does. They are made to feel good about working for the firm.

The second purpose is to let them know about working for the firm. They learn about rules and workers' benefits.

PROGRESSING IN CONSTRUCTION

Those who work hard and are the best-trained progress the quickest.

To progress in a job, both pay and job level may change. Fig. 11-8 is a drawing that shows this concept. A worker can progress in four directions. The first way is up. We most often think of progress in this way. A person who has an *upward progression* will have an increase in pay and job level. Crafters who become supervisors will earn more money. They are in charge of a crew. They will decide things that direct the crew. This is not true for crafters.

Supervisors can progress downward. This may happen because they prefer not to be in charge. A person who cannot handle a crew may be placed at a lower job level and will experience *downward progression.*

Lateral movement results in no change in wages or job level. The transfer of a supervisor to another project is *lateral progression.* The person may have asked for the transfer. Maybe the home office felt the person knew more about the other job.

Separation, or leaving a firm, is the last way to progress. People leave firms for many reasons. The person may feel that progress in some other firm is faster. The person may not be happy working for the firm. The firm may not be happy with the person. That person may quit or may be fired.

What is a Labor Union?

A *labor union* is a group of workers who want to *bargain collectively.* To act collectively means to act as one. To bargain is to discuss and decide with the employer the issues that relate to work. Look at Fig. 11-9. Note some of the issues discussed. Higher wages, better working conditions, and more benefits are three issues that concern workers.

The AFL-CIO (American Federation of Labor-Congress Industrial Organizations) is a national labor group. It is made up of departments representing different kinds of workers. Construction workers are a part of the Building and Construction Trades Department. This department has ten member unions. The carpenters union is one of the largest with over 600,000 members. Construction workers belong to unions called *trade unions.*

Industrial unions represent an entire industry. Both skilled and unskilled workers are members. This kind of union is more common in manufacturing firms.

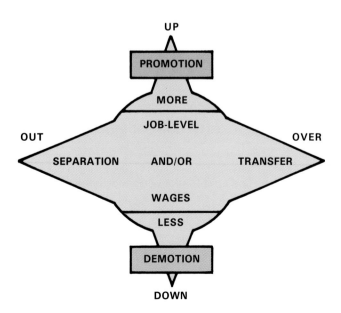

Fig. 11-8. Progress in a job may be your choice or the choice of others. What does this statement mean?

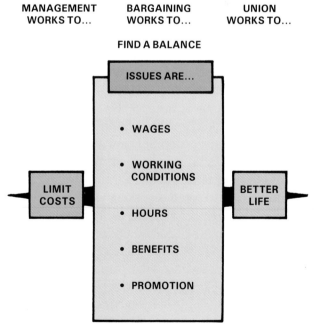

Fig. 11-9. The parties, their goals, and the issues involved in collective bargaining are shown. Try to explain this concept to a friend.

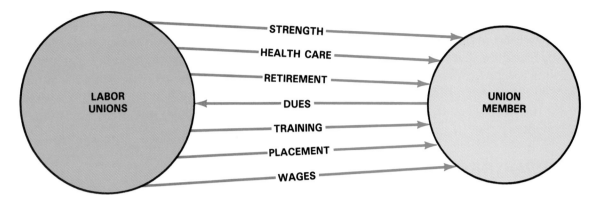

Fig. 11-10 . This drawing shows what union members get for their dues.

Union members, at all levels, elect their leaders. The leader's main job is to organize the workers. A second task is to settle problems between members. The third is to stay informed on labor guidelines. Lastly, they provide training and inform the public about the union.

Why Belong to a Union?

Labor unions are of value to construction workers. One person's voice is not heard. When people join in a group, they gain strength. When people feel they are not treated fairly, they have a *grievance.* They want to be heard. Prepared people speak for workers when a labor contract is being drawn up.

Journeymen (full members) can work on union jobs (jobs where only union members can be hired). The members receive the wages stated in the contract. Union workers can transfer to other local unions.

Members of unions enjoy health care and retirement plans. *Health care plans* apply when a member becomes ill or is injured. Health care plans apply to the workers and their families. *Retirement plans* ensure that workers are paid after they reach the age when they no longer work. While working, part of their wages are put into a fund. Money from the fund is used to pay the workers after they retire.

Training programs increase the chances to progress. Most training in construction occurs on-the-job. This means they learn while they work. New workers are placed with a skilled supervisor or other workers. These people teach the new worker practical skills. Labor unions arrange the training their members need.

Labor unions provide a *placement service.* It works like this. If members are out of work, their names are placed on a list. It is called an unemploy-

ment list. When builders need workers, they contact the union. They ask for the number of workers they need. The union contacts members on the list, and they are sent to the builder. Both members and builders gain from this service. Fig. 11-10 puts the union concept into a picture.

HOW DO YOU BECOME A UNION MEMBER?

Methods of joining a trade union differ. The sequence in Fig. 11-11 is common.

STEP	TASK
1	GET DIPLOMA
2	BE SELECTED
3	FINISH TRIAL PERIOD
4	JOIN UNION
5	PAY INITIATION *
6	GET ON-THE-JOB TRAINING
7	PAY DUES
8	TAKE NIGHT CLASSES
9	BECOME JOURNEYMAN

* Times and fees vary from union to union.

Fig. 11-11. The process for becoming a member of a labor union.

People who want to become members of a trade union, in most cases, must do two things. They will need to finish high school. They will need a sincere interest in the trade.

Next, a joint group of union leaders and builders select the prospects. A 90-day trial period is served.

If they pass this test, they join the union. They are now an *apprentice* (new worker). They pay an initiation fee.

The apprentice works at a lower pay than a *journeyman* (skilled worker). At prescribed times the apprentice gets pay raises. Part of the pay goes to the union as union dues, Fig. 11-12.

A journeyman gives apprentices on-the-job training for a set length of time. A required number of night classes per week are taken during the school year. For example, the carpentry trade requires at least 144 class hours per year. See Fig. 11-13.

When apprentices complete the training program, they become journeymen. This title allows them to enjoy all the benefits the union has won for its members.

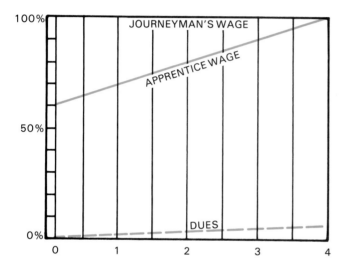

Fig. 11-12. An apprentice is paid less than a journeyman. The pay rises until it reaches a journeyman's wage.

Fig. 11-13. Night classes give the apprentice the ability to use modern construction technology. (United Brotherhood of Carpenters and Joiners)

SUMMARY

Construction workers do site work, build structures, finish structures, and put in mechanical systems. They are hired many times during their working lives. Those who hire workers recruit, select, and induct them. Workers can progress in their jobs. Changes can affect their wages and job level.

Union leaders are chosen by the members. The leaders work to improve the lives of workers. They bargain collectively to do this.

KEY WORDS

All the following words have been used in this chapter. Do you know their meaning?

Apprentice
Bargain collectively
Downward progression
Finish work
Grievance
Health care plan
Inducting
Interview
Job application
Job order
Journeyman
Labor union
Lateral progression
Mechanical work
Placement service
Recruiting
Reference
Retirement plan
Selecting
Separation
Site work
Substructure
Superstructure
Trade union
Training program
Upward progression
Walk-in application

TEST YOUR KNOWLEDGE

Write your answers on a separate sheet of paper. Do not write in this book.

1. Construction workers may change jobs five to eight times a year. True or false?

2. The superintendent is most often a (full/part) time employee of a building firm.
3. The _____ of a small project will hire the workers.
4. Which of the following is not a common step in the hiring process?
   a. Select.
   b. Bargain.
   c. Recruit.
   d. Induct.
5. Often, the _____ hall's agent has a pool of skilled workers.
6. A labor union is a group of workers who want to _____ collectively.
7. The traditional name for a worker who has finished apprenticeship is _____.
8. When workers feel they are not treated fairly, they have a _____.
9. To help you find work, labor unions provide a _____ service.
10. A carpentry apprentice must attend at least _____ hours of night classes per year.

ACTIVITIES

1. Visit a building site. Can you find site work, work on the structure, finish work, and mechanical work? Take pictures of it. Report to the teacher or the class what you found.
2. Visit with a worker. Find out how they were hired. Ask about methods used to recruit, select, and induct them. Make a display board to show how it was done.
3. Phone or visit a union hall. Ask them how you could become a member. Write down the process. Compare it with the one in this chapter.

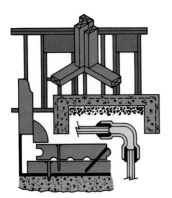

Chapter 12

GETTING MATERIALS AND EQUIPMENT

After studying this chapter, you will be able to:
☐ Tell how lead time affects purchasing.
☐ Define terms such as stock, purchase order, and invoice.
☐ Give reasons for renting or buying equipment.
☐ Compare the methods of obtaining materials, such as lump sum contract and job account.

Construction projects vary greatly, even when they serve the same purpose. Fig. 12-1 shows two bridges serving the same purpose.

Projects may be many miles long. The pipeline in Fig. 12-2 is over 800 miles long. It was hard to protect materials on that project site. For the above reasons, contractors use special methods to buy things.

Builders plan to obtain materials and equipment early, and sign up subcontractors early. Even before the contract is signed, they schedule their buying. They use set methods to lower costs. Records

Fig. 12-2. How do you think the pipe for this project was bought? (Shell Oil Co.)

A

B

Fig. 12-1. Projects built by construction firms differ greatly. A–Rotation is one way to allow ships through. (American Institute of Steel Construction). B–A liftable bridge is very sturdy. Can you find an elevated span? (North Carolina Department of Transportation)

are kept and used to figure taxes. Records also help leaders find ways to save money on future projects.

WHY PURCHASING IS SCHEDULED

As soon as a verbal award of the contract is given, purchasing begins. The notice of award is sent to the builder. It is a paper that says the builder has the contract. This paper serves as the contract until the real contract is drawn up and signed.

Leaders in the builder's firm have a planning meeting, Fig. 12-3. The purchasing agent attends the meeting. A master schedule for the project is made early. A master schedule includes all tasks and will reduce lead time. *Lead time* is the time it takes a firm to deliver its product or service to the site.

Special items require more lead time than standard items. Any special items need to be put into the supplier's work schedule. The 48 in. butterfly valves in Fig. 12-4 were made just for that project. Several weeks lead time was needed.

The lumber in Fig. 12-5 is standard stock. It was bought from a local building supply store. Only a few hours lead time was needed.

Project leaders schedule equipment used to build the project. If the builder owns idle machines, there is little problem. When machines are being used on other projects, there is a bigger problem. One project must wait until another is finished. The time before machines can be used is lead time. The time needed to transport it to the job site must be added.

The machine in Fig. 12-6 was made just for that project. Lead time is reduced when a machine is built right on the site. Lead time is also shorter when standard parts are used to build the machine.

Equipment used in the project is often the prime concern. Some owners request that the vendor build a model of the equipment. The models prove

Fig. 12-4. Lead time was needed to get these parts. How was it used? (American Cast Iron Pipe Co.)

Fig. 12-5. The lumber used in construction comes in standard sizes. It can be bought in any city. (Maine Department of Transportation)

Fig. 12-3. These people are planning the schedule. The schedule will guide the purchasing agent. (Combustion Engineering, Inc.)

Fig. 12-6. Some projects require special machines. This one did. It took only 18 days to design and build. That is fast, but the 18 days must be scheduled. (Morgen Manufacturing Co.)

that the product will work. Fig. 12-7 is a picture of one such model. Making a model may extend the lead time to over a year. Lead time on equipment can delay the entire project.

Early planning helps subcontractors. They must purchase materials and schedule equipment, too. The builder should select subcontractors as soon as possible.

HOW BUILDERS OBTAIN MATERIALS

Builders have at least five options for getting materials. They can get them from stock, buy as needed, buy with lump sum contracts, use a job account, or let a subcontractor buy materials.

From Stock

Material can be gotten from *stock* when a surplus of materials grows in the stockyard. Materials come from other projects. They may be salvage or surplus materials. In either case, they can be used on other projects. Some of the cost of new materials is saved by using materials from stock.

A list of surplus materials is kept. The stock is maintained until a request comes in. The cost of getting and hauling the materials to the site is charged to the job.

Surplus materials are used whenever possible. Care must be taken to assure that the surplus materials meet specifications.

Buy as Needed

The most common way to get materials is to *buy as needed.* This is done so that theft, loss, and damage are reduced. The material is used soon after it is placed on the site. The faster it is used, the less chance there is for loss and damage.

Fig. 12-7. This is a model of an electrostatic precipitator. It cleans the dirt out of smoke at a generating plant. The builder must prove to the owner that it will work. (Dresser Industries, Inc.)

The superintendent fills out a *purchase requisition.* The original and two copies are made. The first copy is kept in the field office. The second goes to the project manager. The original goes to the finance office.

A *purchase order* is made up. Fig. 12-8 is a sample. It is sent to the vendor.

The vendor delivers the materials to the job site. A *bill of lading* (list of items) is checked against the shipment. The material quality is checked. The pieces are counted or measured. The bill of lading is signed.

If a shortage is noted, a claim is filed. Low grade materials are not accepted.

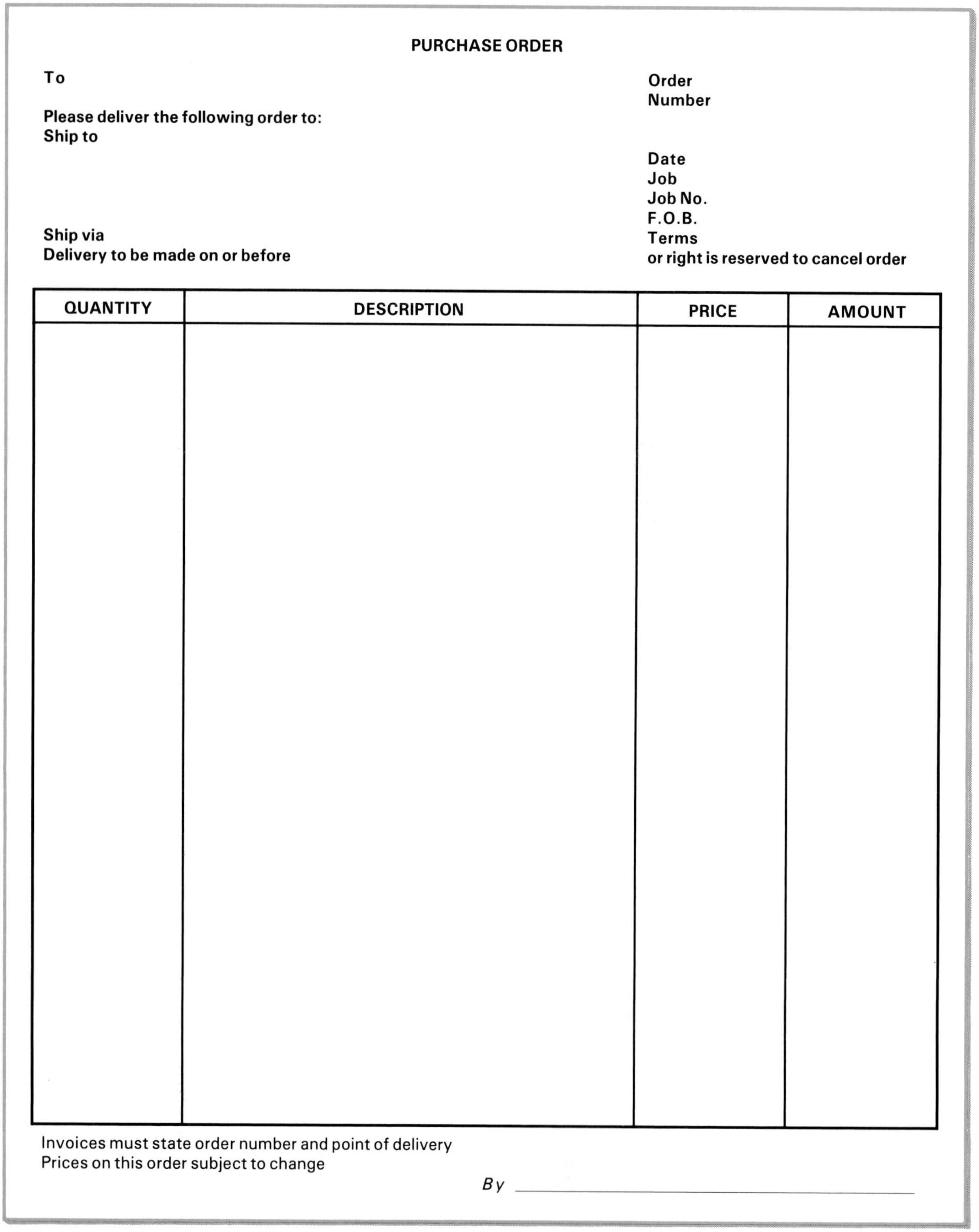

PURCHASE ORDER

To

Please deliver the following order to:
Ship to

Ship via
Delivery to be made on or before

Order
Number

Date
Job
Job No.
F.O.B.
Terms
or right is reserved to cancel order

QUANTITY	DESCRIPTION	PRICE	AMOUNT

Invoices must state order number and point of delivery
Prices on this order subject to change

By _____

Fig. 12-8. Notice the importance of a delivery date.

Larger projects have receiving clerks. Their job is to check the shipments. They sign for each one. Clerks must be reliable and honest.

The vendor sends an invoice to the home office. An *invoice* is the bill to be paid. The accountant matches the purchase order, receiving report, and invoice. They must all show the same goods. When the order is complete, the vendor is paid. A *trade discount* is given if the bill is paid before a stated length of time.

On smaller jobs a supervisor may do the checking. Checking may be thought of as a secondary job. In that case, it may not be done with enough care. Shortages, damage, and lost items may result.

Lump Sum Contracts

With *lump sum contracts,* all materials are sent to or through a central supply. People receive, store, and account for all materials. This method gives better control of the materials. The firm receives orders and sends them to the project. This system shortens lead time for stock items.

First, a *request to purchase* comes from the field office. After approval, the order is shipped out. A *stores ledger* is used to assign materials to each project. A *stores card* is used to keep track of each kind of material.

Job Account

A superintendent has a set amount of money on hand. It is called a *job account.* The money may be in the form of cash, a checking account, or credit. It is used to pay bills that cannot wait. Workers who quit or are fired can be paid. Freight and other charges are paid through this account. A receipt or invoice is used to show how the money was used.

Submaterials

Submaterials are used by the subcontractor. The subcontractors order their own materials. They are in charge of checking, storing, and moving them.

Procedures are about the same as described for the general contractor. The main difference is that the subcontractor is the field office person in charge.

HOW BUILDERS GET EQUIPMENT

A firm often *buys* its own machines for big jobs. This happens more often when the work is common to the firm. Fig. 12-9 shows a machine that the contractor owns. Firms buy equipment when there is enough lead time. It takes time to buy it, make it, and ship it. This lead time must fit into the construction schedule.

Equipment is *rented* when it will be used for a short time. Machines that are needed right away are often rented. The machine shown in Fig. 12-10 is rented. Can you guess why? It may cost less to rent equipment you need right away than to wait for a purchase. The cost of a delay in the schedule may be more than the rental fee.

Fig. 12-9. This machine grades the roadbed for paving. The contractor owns the machine. (CMI)

Fig. 12-10. What about the machine that transfers the concrete to the footings? Is it rented or owned by the contractor? (MORGEN Manufacturing Company)

SUMMARY

Keeping on schedule is not easy. It is impossible if materials and equipment are not on the site. It is the purchasing agent's job to select the best vendors. This does not always mean the one with the lowest price. Lead time is crucial in choosing a vendor.

Materials for a project can be gotten from stock, as needed, by a lump sum contract, or from the job account. Equipment is either owned or rented by the firm.

KEY WORDS

All the following words have been used in this chapter. Do you know their meaning?

Bill of lading
Buy as needed
Invoice
Job account
Lead time
Lump sum contract
Purchase order
Purchase requisition
Request to purchase
Stock
Stores card
Stores ledger
Submaterial
Trade discount

TEST YOUR KNOWLEDGE

Write your answers on a separate sheet of paper. Do not write in this book.

1. Theft, loss, and damage are less if a contractor buys as _____.
2. The time it takes a firm to deliver its product or service to the site is called _____ _____.
3. Select subcontractors as soon as possible, because they need time to arrange labor and get _____.
4. Surplus materials can be used if the builder decides to get materials from _____.
5. What is a vendor's request (or bill) to be paid for the products delivered?
   a. Purchase requisition.
   b. Purchase order.
   c. Invoice.
   d. Bill of lading.
   e. Receiving report.
   f. Claim.
6. With _____ sum contracts, all materials are sent to or through a central supply.
7. Machines that are needed right away are often (bought/rented).
8. The cost of a delay in the schedule may be more than an equipment rental fee. True or false?

ACTIVITIES

1. Measure your front or back step for a hand rail. Get prices for the railing from two or more vendors in your town.
2. Ask each vendor if they deliver. If they do, ask what it will cost.
3. Go to their stores. Look closely at each railing. Are they made the same way? Do you think they will all stand up over time?
4. Are they fastened to the step in the same way?
5. Would having the railing made for you affect your construction schedule? In what way?
6. Do you have the equipment you need to install the railing? What would it cost to rent the equipment? What will it cost to buy it? Which method would you use?
7. Choose the best vendor for you.

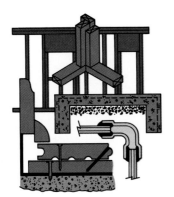

Chapter 13

CONSTRUCTION MATERIALS

After studying this chapter, you will be able to:
☐ List the materials that make up concrete.
☐ Discuss concrete workability and curing time.
☐ Define lumber terms and grades.
☐ Describe plywood, laminated wood, and pressure treated wood.
☐ List advantages of using metal for framework and reinforcement.
☐ Explain how brick and block masonary are placed and finished.
☐ Give uses for types of glass panels.

Suppose you want to build a structure to hold two trash cans. You want to keep them out of sight, yet easy to reach. You want the structure to be strong enough to last and to look good. You have many choices of materials. Concrete can be used for the bottom. The fence can be made of wood, metal, bricks, or concrete blocks. The materials you choose will depend on their properties and cost. This chapter describes five common construction materials. They are concrete, wood, metal, masonry, and glass. Can you find these materials in Fig. 13-1?

CONCRETE

Concrete is one of the most important building materials. It is used in almost every type and size of structure. In buildings it is used for footings, foundations, and walls. It is used for flat structures of all kinds, like roadways and floors. Fig. 13-2 shows a range of uses for concrete.

Ingredients

Concrete is made of two parts: cement paste and aggregate. The *cement paste* is portland cement and

Fig. 13-1. What construction materials can you find in and around this home? (Pella/Rolscreen Co.)

water. The *aggregate* consists of fine sand (1/4 in. in diameter and less) and stones over 1/4 in. in diameter. The cement paste binds the aggregate together and fills the spaces between particles. See Fig. 13-3. A chemical reaction between the water and the cement sets (hardens) the concrete.

Admixtures are special chemicals added to concrete. Some chemicals make the concrete set faster. Others will lengthen the setting time.

Concrete with tiny air bubbles in it is called *air-entrained concrete.* An admixture causes the tiny bubbles to form. Air-entrained concrete is easier to work, and it resists cracking from freezing and thawing.

A

B   C

Fig. 13-2. How many uses of concrete can you find? A—Northern highways must withstand ice and salt. (HNTB Engineers) B—Concrete left in water becomes very strong. (The Manitowoc Co., Inc.) C—Concrete is more stable than asphalt or other materials. (Raymond International, Inc.)

Fig. 13-3. Notice the different sized aggregate in this concrete sample. The space between the sand and gravel is filled with the cement paste.

Reinforcing rods, called rebar, are used to strengthen concrete. They are made of steel. The surface of each rod is often rough or coated to make the concrete hold better. Fig. 13-4 shows some forms of reinforcing rods.

Glass, plastic, and wire fibers are sometimes added to concrete. They reduce the number of small cracks that appear in concrete. You can see why it works if you look at Fig. 13-5.

Properties

Material properties determine how materials should be used. Often more than one property is considered before a material is selected.

The main properties engineers consider in concrete are strength, watertightness, durability, and workability. Concrete must have the strength to carry heavy loads and must not wear away. Fig. 13-6 shows a project that requires strong concrete.

Concrete that is exposed to water must be watertight. A concrete dam should not leak, nor should water be able to soak into concrete and freeze. This may cause it to break.

Concrete is used where it is expected to have a long life with low upkeep. Roadways, dams, and other structures are built to last. Concrete, therefore, must be durable.

Workable concrete flows in and through steel rods and into the corners of forms. See Fig. 13-7. If concrete is not workable, gaps around rods and

Fig. 13-4. Steel rods are used in the columns and floor of this foundation. (Bob Dale)

holes in corners of forms appear. The first problem weakens the concrete, and the second looks bad or unfinished.

An engineer can get a suitable quality by changing the materials used. A *suitable quality* is one that gives the intended service in use. That means it will last the planned lifetime of the project, and is easy to place.

Fig. 13-5. Small steel wires were added to the concrete mixture. The concrete resists cracking. (Ribbon Technology Corp.)

Fig. 13-6. A concrete bridge must support heavy loads. (Michigan Department of Transportation)

Fig. 13-7. Concrete used on this bridge deck must be workable. It has to flow in and around two or more layers of steel. (CMI)

The mixture of materials determines the quality of concrete. The amount of water in the mix determines many of the properties. Water makes concrete easy to work. But, extra water will reduce the strength, durability, and watertightness of concrete.

Curing the concrete also affects quality. Proper curing requires time, temperatures above freezing, and moisture. Fig. 13-8 shows how time and moisture affect strength of concrete.

In cold weather, fresh concrete must be kept warm. This can be done by warming the materials and covering the completed job. It is kept moist by sprinkling and by using various coverings.

WORKING WITH CONCRETE

Concrete work requires four steps. These are: preparing, placing, finishing, and curing.

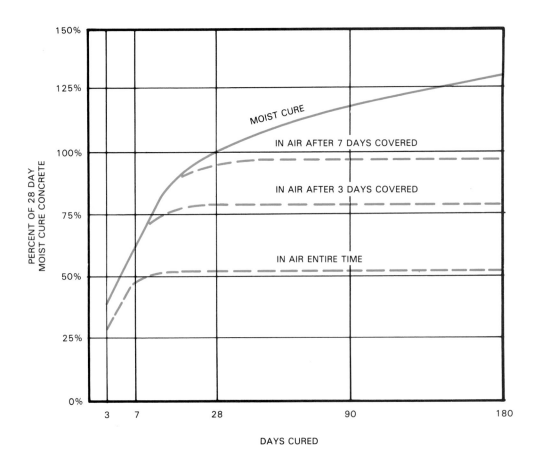

Fig. 13-8. Notice that the longer concrete is kept moist, the stronger it becomes.

Preparing

Forms and subgrade for concrete should be properly prepared. See Fig. 13-9. Forms usually are leveled and well braced. They should be clean and tight so fresh concrete is contained. Reinforcing should be in place and up off subgrade the correct distance.

The *subgrade,* the leveled suface on which the foundation is placed, should be at the correct elevation. Porous material should be compacted and moist. A moist subgrade does not soak up the water from fresh concrete.

Placing

Concrete should be placed as near as possible to its final location. See Fig. 13-10. Moving fresh concrete is hard work and may cause the aggregate to

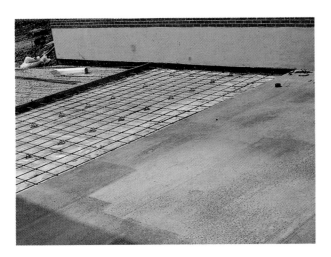

Fig. 13-9. You can see a firm subgrade, some steel, and part of a form. An expansion joint is made of soft material. It prevents concrete from buckling when it gets hot.

A

B

C

D

Fig. 13-10. Methods for moving concrete to its place of use. A–Conveyors. (Morgen Mfg. Co.) B–Pumps. (Morgen Mfg. Co.) C–Buckets. (DYK Prestressed Tanks, Inc.) D–Chute. (The Burke Co.)

segregate. This means the coarse and fine particles tend to separate.

Finishing

Concrete finishing begins with *screeding.* The surface is struck off with a straightedge. The straightedge is moved back and forth across the top of the forms. Darbying is done next. *Darbying* levels and smooths the concrete. Screeding and darbying are being done in Fig. 13-11.

Edging and jointing follows. *Edging* is rounding the edges over to make them stronger. With *jointing,* joints are made to control cracking.

Floating is done to push down large rocks and prepare the surface for the final finish. A 3 to 4 ft. long board or metal plate is used.

Final finishing is done with steel trowels, brooms, and other tools. Steel trowels are used to produce a smooth finish. A broom is used to produce a nonslip finish for walkways and roadways. The walkway in Fig. 13-12 is being finished with a broom. All of the finishing jobs can be done with one machine in Fig. 13-13.

Curing

Curing, the setting of concrete, is often done with moist conditions by sprinkling, ponding, or by using a plastic sheet or burlap as a cover. Sometimes a waterproof coating is sprayed on the surface. Forms should be moved early to allow patching as soon as possible. However, the concrete must not be too fragile when forms are removed.

Fig. 13-11. Fresh concrete is being screeded.

Fig. 13-12. A final finish is put on after the concrete begins to set. A broom is being used to put a nonskid surface on this sidewalk.

Fig. 13-13. Machines do much of the hard work and reduce the number of workers needed to work concrete. (U.S. Department of Transportation, Federal Highway Administration)

WOOD MATERIALS

Wood is a renewable resource. Wise forest management assures a lasting supply. Aside from its availability, it is very suitable for construction. Wood has high strength for its weight. It is easy to work with wood. If protected, wood lasts a long time. Some woods are highly prized for their beauty. These properties are important in lumber, plywood, laminated timbers, and other products.

Lumber

Lumber is a product of the saw and planing mills. *Yard lumber* comes in the forms of boards, dimension stock, and timbers. Boards are used for sheathing, siding, flooring, trim, and paneling. What forms of yard lumber are used for the house, porch, steps, and railing shown in Fig. 13-14? *Dimension lumber* forms parts of a wood frame for a house. Posts and beams are made of timbers.

Lumber is produced in many grades. *Grading* is used to describe the quality of lumber. Quality is determined by appearance and strength.

Finish grades are determined mostly by appearance. This is affected by the number and size of defects. The wider the board and the better it looks, the more it costs. Finished lumber is used where it will be seen. High grade lumber is being made into moldings in Fig. 13-15.

Structural grades are influenced by the size and location of defects. These defects affect the strength and straightness of lumber. The longer, stronger, and straighter the lumber, the more it costs. You can see how important it is to have straight lumber in Fig. 13-16.

Fig. 13-15. This lumber has no defects that you can see. It is a high grade because it looks good. (Jordan Millwork Co.)

Fig. 13-14. Can you find the boards, dimension lumber, and timbers in the house, porch, steps, and railing?

Fig. 13-16. Why must the dimension lumber in this picture be strong and straight? A roof using trusses can be finished in one day.

Plywood

Plywood is used as support and to decorate the structure. Structural plywood is used as wall sheathing, roof decks, and subfloors. *Plywood* is made of thin crossbanded layers (with alternate grain at 90° angles). The layers are glued together into panels.

Structural plywood is used where strength is desired. Two basic types are produced. Interior plywood is made with glue that is moisture resistant but not waterproof. Exterior glues are waterproof. The layers stay bonded even when they are wet and dried many times.

Decorative plywood may have designs cut in it with special machines. Sheets may have a veneer of hardwood. They may also have vinyl coverings or printed patterns.

Each type of plywood is made with several **appearance grades.** The grades range from N to D. The chart in Fig. 13-17 describes the grades.

Plywood comes in panels. They are most often 4 ft. wide and 8 ft. long. Other sizes can be purchased. The other sizes usually need to be ordered.

Laminated Timbers

It is not always possible to find trees the size and shape you need for the project. Laminated timbers were used to build the structure in Fig. 13-18. Arches, poles, and beams are laminated. **Laminated timbers** are made from the lumber that is glued together. The glue is stronger than the wood itself.

Other Forms

Small defective trees can be reduced to chips. The chips are glued together and cut into 4' x 8' sheets of different thicknesses.

Particleboard is made from small chips glued together. The smallest chips are used on the outside of the board so the surface will be smooth. Larger chips are used in the core for strength. Particleboard is used more for its smooth surface than for its strength.

Oriented strand board (OSB), sometimes referred to as flakeboard, is nearly as strong as plywood but less costly. Large chips or flakes of wood are aligned and glued together. This makes the board stronger. It is used in place of plywood for structural members of buildings. OSB is being used in Fig. 13-19.

Hardwood is made from specially processed wood fibers. The panels are usually 1/8″ to 1/2″ thick. Some are glassy smooth on one or both sides. Hardboard can be made waterproof and used for

GRADE	DEFECTS PERMITTED	AVAILABILITY
N	No open defects All heartwood or sapwood	Special order
A	Smooth paint grade	Standard
B	Solid with neat patches Lowest grade on exterior	Standard
C	1 1/2″ tight knots 1″ knotholes	Standard
D	Lowest grade on interior 2 1/2″ knotholes 1″ wide splits 1″ wide splits 2 1/2″ pitch pockets	Standard

Fig 13-17. Structural plywood is graded on the basis of the face veneers. A-C is a grade. Can you explain it?

Fig. 13-18. It would be hard to find 10 trees the size and shape to build these rafters. When you laminate framing members, you can make them any size or shape you please. (Standard Structures, Inc.)

Fig. 13-19. Oriented strand board (OSB) is used at the corners to add strength.

A

B

C

Fig. 13-20. Fiberboard can be used in many ways.
A–On a roof. B–As paneling. C–For siding.
(Masonite Corp.)

curved concrete forms. Underlayment is another use for smooth hardboard. Underlayment is a smooth surface placed beneath vinyl floor coverings. Most of the textured hardboard is used for siding, wall paneling, and roofing. Fig. 13-20 shows wood fiber products used for roofs, paneling, and siding.

Pressure Treated Wood

When kept dry and free from insects, wood will last for many years. Serious decay and insect problems can ruin wood structures. Decay and mold occur when wood is damp, has oxygen, and is at mild temperatures. Insects bore into the wood and feed on it. Termites are the biggest problem.

Preservative treatments protect wood against decay, mold, and insects. *Pressure treatments* force the chemicals deep into the wood. Pressure treated wood can be used underground. It is used as a foundation in Fig. 13-21. Wood that is not pressure treated is used above ground only.

Fig. 13-21. Pressure treated plywood and lumber are suitable for use underground. (Hickson Corp.)

METAL MATERIALS

Many metals are used in construction. Steel, aluminum, and copper are used most often.

Metal products are produced in many ways. The most common methods are casting, rolling, extruding, and assembling. Cast products are made in foundries. Casting is used for complex shapes. *Casting* is done by pouring molten metal into sand molds. The grates used in Fig. 13-22 were constructed that way.

Rolling is acomplished by shaping sheets, strips, and blocks of metal between shaped rollers. Sheets, structural shapes, and bars are made by rolling. Sheets are used to cover the exteriors of structures.

Fig. 13-22. Cast steel grates are made to cover soil around plantings and let water and nutrients get to the tree. (Neenah Foundry Co.)

Structural shapes are most often used for frames in a structure. They are shaped like the letters, H, I, L, and V. Rods are made in many shapes. See Fig. 13-23. A very common shape is round. It is used to reinforce concrete.

Fig. 13-23. Rolled structural shapes and asssembled members are used in this structure. Can you tell which are rolled and which are assembled? (Stran Buildings)

Extruded metals are very useful. Aluminum is extruded easier than most other metals. *Extruding* is like squeezing toothpaste onto your toothbrush. The toothpaste is shaped by the opening. Frames around storm doors, storm windows, and entries to buildings are often extruded.

Standard shapes are not always the best for parts of some structures. Sheets, plates, and rods are cut and later assembled into members for a structure. Several parts in Fig. 13-23 were made through the process of *assembling.*

How Metals are Used

Steel is used for frames. It costs less than other metals and is very strong. Coatings can be put on steel to keep it from rusting. Paint and galvanizing are the most common. *Galvanizing* is a thin coat of zinc. Zinc is a metal that does not rust.

Aluminum is used outside. It looks nice and does not corrode easily. It can be coated in many colors.

Copper is easy to work and join. It does not corrode beyond a thin layer. It is used for plumbing pipes and roofing.

MASONRY MATERIALS

Clay and concrete masonry materials are common in construction. Clay masonry is one of the oldest manufactured building materials. *Clay masonry* is made from fired clay units and mortar. Clay units are both solid (bricks) and hollow (tile).

Concrete masonry is made from specially mixed concrete. They come in both solid (brick) and hollow (block) forms. See Fig. 13-24.

Masonry is used mostly for walls. Masonry exterior walls are strong, watertight, and durable. They also resist the transfer of heat, sound, and fire. Brick walls blend well with other materials. When masonry is exposed to view, the color, texture, and pattern are important.

Mortar bonds the individual units into a structure. Mortar is made of cement, lime, sand, and water.

Building Masonry Projects

Laying masonry units follow a procedure. Bricklayers lay out the job, lay up corners, lay between corners, and finish joints (tool them). These steps are shown in Fig. 13-25.

Fig. 13-24. Concrete block are placed in mortar 3/4 in. thick, which is squeezed down to 3/8 in. (Ingersoll-Rand Co.)

Laying out

Laying out means masonry units are spaced out in their planned positions. No mortar is used. This step determines how to space and cut the units. The first course (layer) is laid. It is made level, straight, and at the correct height.

Fig. 13-25. Brick and block are laid into walls in much the same way. For both brick and block, masons lay out the job, build leads, and do the ''lay-betweens'' (between corners). (Mary Robey)

Laying up corners

Corners of the wall are built next. They are called *leads.* They are laid straight, square, level, and plumb. Mortar joints are usually 3/8 in. thick. The horizontal joints are called *bed joints. Head joints* are on the ends of the units.

Laying between corners

When filling in the walls between the corners, a *mason's line* is used. It is stretched between the corners. Mortar is spread on the lower course (layer). The ends of the unit are buttered (mortar put on it). It is tapped into place. Care is taken to keep them level and lay them to the line. The last unit to be placed in the course is called the **closer unit.**

Tooling joints

Mortar joints are finished by tooling. *Tooling* compacts and shapes the mortar and seals each side of the joint. Proper tooling is done after the mortar is "thumbprint" hard. Fig. 13-26 shows nine ways to tool mortar joints.

GLASS

The use of glass has increased in recent years. Today, all-glass buildings or those mostly of glass are common. See Fig. 13-27.

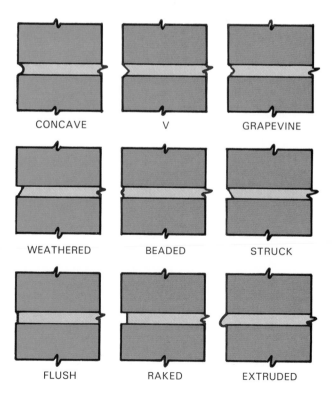

Fig. 13-26. Mortar joints are finished in many ways. Here are nine of them.

Fig. 13-27. The glass used in this building is both tinted and insulated. (Libbey-Owens-Ford Co.)

Most glass used in construction is *flat glass.* It can be either clear, tinted, or opaque. Clear glass provides true vision. Tinted glass reduces glare and absorbs solar energy. Opaque glass lets light in, but prevents clear vision.

Regular glass comes in thicknesses of 1/8 to 1/4 in. It is used for small windows and small mirrors, Fig. 13-28.

Heavy glass is 5/16 to 7/8 in. in thickness. It is used for large store fronts and doors.

Pattern glass has a pattern on the surface. Wired glass has a wire mesh rolled into it. The wire holds the glass together under low-level impact.

Tinted glass is mostly made in green and bronze colors. Tinted glass is used where some of the heat and light are to be kept out of a building.

Reflective glass will reduce the solar heat coming into a building. This glass has a coating bonded onto the surface of the glass. A wide range of colors is produced.

Tempered glass is heat treated. Heat treating makes it harder to break. When it does break, the pieces are small and harmless. Tempered glass is

Fig. 13-28. Glass for small windows is about 1/8 in. thick. (Pella/Rolscreen Co.)

used in doors and windows where people may come in contact with it. This type is also called safety glass.

Composite glass is made of two or more layers of glass. Composite glass includes insulated glass and laminated glass.

Two layers of glass with dry air between them is called **insulated glass.** The glass panes in Fig. 13-29 are insulated panels.

Fig. 13-29. Double glazed insulated glass holds in twice as much heat as single glazing. (Pella/Rolscreen Co.)

Glass block provide the highest insulation value. Each block is sealed to form a vacuum inside.

Laminated glass consists of two or more layers of glass with a tough plastic sheet between them. The glass and plastic are bonded to form a single unit. Laminated glass is strong. When it breaks, most of the pieces stick to the plastic core. The glass panels in Fig. 13-27 were laminated.

SUMMARY

Construction projects consume huge amounts of materials. Construction materials must be reasonably priced, be strong, look nice, and last a long time. Concrete, wood, metals, masonry, and glass have those qualities. As a result, these materials are used in construction the world over.

KEY WORDS

All the following words have been used in this chapter. Do you know their meaning?

Admixture
Aggregate
Air-entrained concrete
Appearance grade
Assembling
Casting
Cement paste
Clay masonry
Closer unit
Composite glass
Concrete
Concrete masonry
Curing
Darbying
Dimension lumber
Edging
Extruding
Finish grade
Flat glass
Floating
Galvanizing
Glass block
Grading
Hardboard
Insulated glass
Jointing
Laminated glass
Laminated timber
Laying out
Leads
Mortar
Oriented strand board
Particle board
Plywood
Preservative treatment
Pressure treatment
Rolling
Screeding
Segregate
Structural grade
Subgrade
Suitable quality
Tempered glass
Tooling
Yard lumber

TEST YOUR KNOWLEDGE

Write your answers on a separate sheet of paper. Do not write in this book.

1. Concrete is made of two parts, cement paste and _____.
2. A chemical _____ between the water and cement hardens the concrete.
3. A porous subgrade under fresh concrete should be compacted and _____.
4. The following are steps for working with concrete. Arrange them in the proper order.
   a. Floating.
   b. Derbying.
   c. Jointing.
   d. Screeding.
   e. Edging.
5. The wood frame for a house is built with (yard/shop/dimension) lumber.
6. Plywood is made of thin layers with alternate grain at _____ angles.
7. Wood strips or sections are _____ to form large beams and arches with curved shapes.
8. To prevent decay and insect damage, wood is _____ treated.
9. List three ways metal shapes are formed.
10. Galvanizing is a thin coating of _____.
    a. lead.
    b. copper.
    c. silicon.
    d. zinc.
    e. tin.
11. What common silvery metal does not corrode easily?
12. Mortar is made of cement, _____, sand, and water.
13. The starting corners of a masonry wall are called _____.
    a. lay-betweens.
    b. head joints.
    c. leads.
    d. bed joints.
14. Mortar joints in brick or block work are finished by _____ them.
15. A type of glass that resists breaking is called _____ glass.
16. Two or more glass layers which have a tough plastic sheet between them make up _____ glass.

ACTIVITIES

1. Design a new mailbox support out of concrete, out of wood, out of metal, out of masonry, and out of glass. How similar do they look? Do they cost the same? Which one(s) could you make? Which one would you choose to build? Which materials can you buy in your town?
2. At a hardware store, find out prices of concrete in small bags. Do the same for patching concrete and for mortar. Are there any instructions that you did not know about? What repair project could be easily done on your home driveway, sidewalk, or gateway?

This redwood lumber is separated with spacers called "stickers" to promote more efficient air drying. Wood seasoned in the open air should be allowed to dry for at least a month. (California Redwood Assoc.)

Section 5
BUILDING THE STRUCTURE

Chapter 14
PREPARING TO BUILD

After studying this chapter, you will be able to:
☐ Explain when a building permit is or is not required in construction.
☐ Describe both the salvaging and the wrecking operations.
☐ Describe ways to remove brush using chains.
☐ Discuss surveying using centerlines and baselines.
☐ List requirements for utilities, temporary shelter, and access to the site.
☐ Suggest ways to make the work site safe.

The builder must get a permit before building begins. Next, the site is prepared. Extra structures and plant growth are cleared. Roads and utilities are arranged. Stakes are set for building lines. Excavation on the site is started for building.

GETTING A BUILDING PERMIT

People must have permission to use the land. This permission is called a *building permit*. Permits are needed to build, move, or demolish (destroy) structures. They are also needed to enlarge, repair, and to convert a building. Small normal repair to a structure does not require a permit. What is a "normal repair" and what is a "major repair" are explained in this chapter. To *convert* is to change the use of a building. The rules vary from place to place.

It is the builder's job to get the permit. The future project must meet certain standards. When the standards are met, a building permit is given. The standards are stated in the *building code*.

Getting a building permit for a new structure requires complete plans. See Fig. 14-1. Drawings and facts about soil, utilities, materials, and methods are needed. It is easy to get a permit to repair or move a structure. Changing the location of a building on a site is easy, too. A drawing to show where the building will be placed is enough.

The fee paid for building permits varies from place to place. The fee pays the cost to inspect the plans and to issue the permit. As a rule, the larger the project, the more a permit costs.

The building inspector checks how the structure will be used. How land can be used is stated in the *zoning code*.

The type of construction is checked next. Building methods and materials are studied. A structure needs to be set away from the boundaries. The required distance is stated in the zoning code.

Some cities control the designs of buildings. Consider places where earthquakes are common. Cities in these places limit the height of buildings and state how they are to be built.

Fig. 14-1. Building inspectors check the plans before construction begins.
(International Conference of Building Officials)

When a permit is given, a copy is sent to the tax assessor's office. The *tax assessor* sets the value of the property after construction. If the value of the property rises, the city will increase the taxes. If a structure is removed or taken down, the property value falls. Therefore, less taxes will be paid on the property in the future.

A builder must display the permit. Work can be stopped if there is no permit or construction does not follow the building code. In that case, a *stop work notice* is given, Fig. 14-2. Work is not to go on until changes are made.

Fig. 14-2. A stop work notice is legal and binding. (Muncie Newspapers, Inc.)

CLEARING THE SITE

The builder needs to know what features of the site need to be cleared. These features may be structures or trees and brush. If these features are in the way, they are cleared. See Fig. 14-3.

The crew that clears the site can either salvage or demolish the features. Fig. 14-4 shows how to take care of some site features.

Methods of Clearing Sites

There are five ways to clear a structure from a site. These are: wrecking, breaking, blasting, salvaging, and moving.

When machines are used to demolish a structure, it is called *wrecking.* The structure in Fig. 14-5 is being wrecked. A steel wrecking ball is often used.

Fig. 14-3. Note what had to be cleared from these sites. How was it done? A—The wear surface on a highway. (Ingersoll-Rand Company) B—Trees were removed. (Shell Oil Company) C—A sidewalk is being removed. (Caterpillar Inc.)

A

B

C

Fig.14-4. Both structures and plant growth are demolished and salvaged. A—Disgarding rubble. (Caterpillar Inc.) B—Clearing trees and bush. (Asplundh) C—Underground storage tanks are usually removed. (Western Technologies, Inc.)

Fig. 14-5. Wrecking is the method chosen for small buildings.

The heavy ball hangs from a crane cable and the crane operator swings the ball against the building. *Rubble,* the broken fragments of the structure, is loaded onto trucks and hauled away.

In Fig. 14-6, the *breaking* method is being used. Air hammers do the work. They break concrete and rocks into smaller pieces. Then the pieces are loaded into trucks or trash containers and dumped.

Blasting uses explosives to destroy a structure. Blasting can reduce a large structure to a pile of rubble. It takes only a few seconds once the explosives are placed. See Fig. 14-7.

A fourth way to remove features is to *salvage,* or save them. The steel in the bridge in Fig. 14-6 was salvaged. Cutting was used to make the pieces easy to handle. A torch was used to do the cutting.

Some salvaged materials are reused in other structures. The concrete rubble in Fig. 14-8 was an old highway. The machine in the background crushed it and removed the steel reinforcing rods.

Fig. 14-6. An air hammer is breaking the concrete. The steel will be salvaged.

Fig. 14-7. Blasting was used to demolish this structure. Windows, plumbing fixtures, and doors were salvaged before it was blasted. (Cleveland Demolition Co.)

Fig. 14-9. This house will continue to be used on another site. It would be wasteful to demolish it. (Muncie Newspapers, Inc.)

Fig. 14-8. The crusher in the background makes roadbed material from old concrete. The steel reinforcement is removed and sold.

The crushed concrete was made into a new roadbed. The steel was sold as scrap. It will be melted down and used in other products.

Moving is sometimes used with good structures. They are placed on wheels. A truck tractor pulls the structure to a new site. This practice saves a lot of materials and labor. The house in Fig. 14-9 is being moved.

Clearing Trees and Brush

Trees and brush are sometimes in the way. The sites in Fig. 14-10 required the clearing of trees and brush.

A

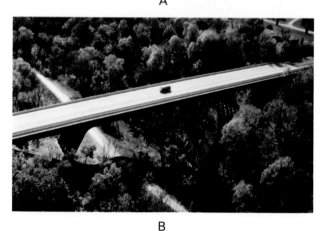

B

Fig. 14-10. Trees to be cleared can be sold. A—The right kind of trees are used for posts. (Caterpillar Inc.) B—Trees from this site may have been used for lumber or firewood. (U.S. Department of Transportation, Federal Highway Administration)

Trees that have value are salvaged. They are cut down and the large logs are made into lumber or veneer. The larger branches can be cut into firewood. A special machine is being used in Fig. 14-11 to remove the stump.

Small trees and brush are destroyed. They can be pushed into piles. Crawler tractors with dozer blades are used. Fig. 14-12 shows methods used to pile brush.

Brush can be knocked over with a heavy cable or chain. Fig. 14-13 shows how this is done.

Next, the workers dispose of the piles of brush. They can be burned, buried, or hauled away. The method used depends on many things. Is burning allowed? Buried wood decays and the soil over the hole will settle. Will that cause a problem? How far must it be hauled? Will the brush need to be chipped? Chipped brush is reduced to small pieces. See Fig. 14-14.

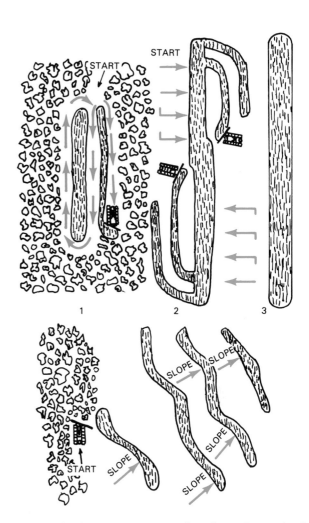

Fig. 14-11. This machine can remove a stump in a matter of minutes. (Vermeer Manufacturing Co.)

LOCATING A STRUCTURE

There are two methods used to locate structures. The first uses a centerline as a guide. The second uses a baseline for control points.

The centerline method is used for highways. The baseline method is used for buildings.

Using a Centerline

Roadways, tunnels, shafts, and piping systems are placed on centerlines. (A shaft is a vertical open-

Fig. 14-12. Here are two ways brush can be pushed into piles. It will be disposed of by burning, burying, or dumping.

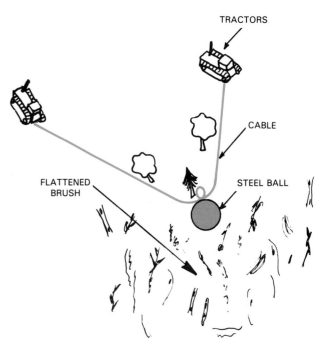

Fig. 14-13. Small trees and brush are knocked over with special chains and cables.

Preparing to Build   151

Fig. 14-14. Brush is chipped into small pieces. It can be hauled away, burned, or used as mulch. (Vermeer Manufacturing, Co.)

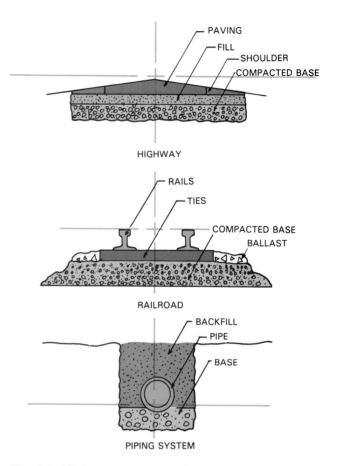

Fig. 14-15. Long narrow projects are laid out on a centerline. A centerline gives horizontal and vertical control.

ing.) A *centerline* passes through the middle of the project. It describes the horizontal and vertical placement of the project. Horizontal means side to side. Vertical means up and down. Fig. 14-15 shows a project that would use a centerline.

A highway is an example that shows how a centerline is used. Fig. 14-16 shows you the process. The first step is to find landmarks shown on the drawing. *Landmarks* are fixed objects that mark a boundary or fix a location. From these points the survey party can determine the direction of the highway.

Highway projects have starting points along the route. In this way, more than one survey party can work at one time.

In the third step, stakes are placed every 50 to 100 ft. along the centerline. The fourth step is to mark the road's width. Stakes are set on both sides of the centerline.

Elevations, or heights, are measured in the fifth step. Leveling techniques described in Chapter 6 are used.

Elevations are marked on the stakes used for horizontal control. The actual land elevations will differ from those planned for the road. The high spots are cut away. The low spots are filled. The process is called *cut and fill.*

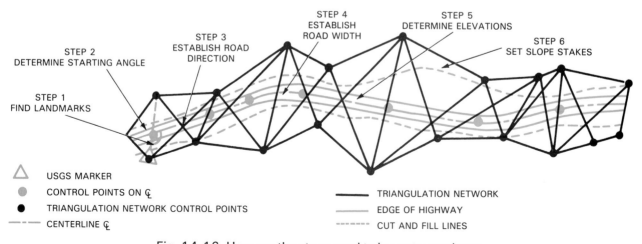

Fig. 14-16. Here are the steps used to lay out a roadway.

Fig. 14-17. A road cut is being made in the background. The excavated earth and rock is used as fill in the foreground.
(North Carolina State Highway Commission)

Fig. 14-17 shows both cuts and fills. The amounts of cut and fill are marked on each stake. The stakes guide the workers when they grade the road bed.

The sixth step is to place slope stakes. *Slope stakes* are placed to show the width at the top of the cut. The width at the base of a fill is also shown. Fig. 14-18 shows a range of cut and fill profiles. Stakes are placed one to two feet away from the cut or fill line. That way, the stakes do not get covered up or moved.

All measurements are carefully recorded, Fig. 14-19. These records are used to check, retrace, or reset stakes.

The road stakes are moved or covered while earthwork is being done. Stakes are replaced after the cut and fill work is complete. New stakes are put on each side of the road. They are placed away from where machines run.

The tops of the stakes are set even with the top of the finished road. Stakes used to guide the paving machines are measured from these stakes. Sensors on the machine guide it. Both the direction of travel and height are controlled. Look at Fig. 14-20. You can see the line, stakes, and sensors.

Baselines for Buildings

A *baseline* is used to place a building. A baseline may be on the axis (line at center) or off to one side. The *center baseline* passes through the middle of the structure. An *offset baseline* is set a distance

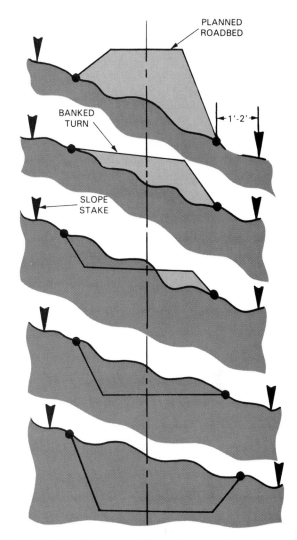

Fig. 14-18. Here are five places where slope is planned. Rock requires less slope. Sand requires more. Refer back to Fig. 8-10.

Fig. 14-19. Careful notes are taken of surveys. They are used to replace and check stakes. (Michigan Department of State Highways and Transportation.)

Preparing to Build 153

Fig. 14-20. A tight string and sensors control this grading machine. Both direction and depth are controlled. (CMI)

As you read the following steps, refer to each number in the drawing:

1. Locate the survey markers that now exist. They may be boundary lines, streets, curbs, or other buildings. These markers are described by the person who designed the building.
2. Locate a centerline of the structure.
3. Clearly mark the centerline.
4. Measure from the centerline to place markers for the offset baseline. These markers should be braced, flagged, and shielded.
5. Locate the corners of the building. The distance from the baseline to each point will be shown on the plot plan. Drive stakes at each corner. The exact corner point is marked on the top of the stake. Nails are used and the height of the nail shows the elevation.

These stakes will be disturbed when digging begins. Therefore, points out away from the work must be set. These points are called reference points. They are located at the number 6 in the drawing.

Reference points are placed on batter boards. Fig. 14-22 shows the parts and how they are built.

from the building. This line is parallel to the centerline. Parallel means that the two lines will never meet.

Follow the process used in Fig. 14-21 to locate a building. The steps are numbered in the drawing.

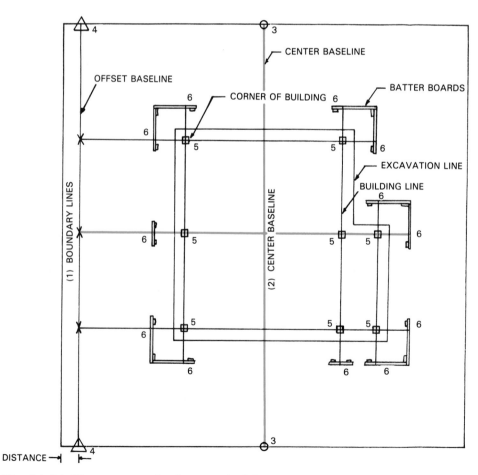

Fig. 14-21. The steps used to lay out a building are shown. Both a center axis and baseline are drawn in.

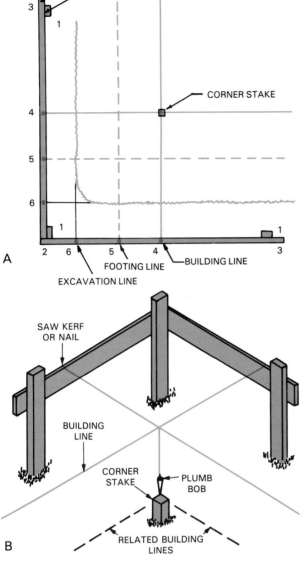

A

B

Fig. 14-22. Using batter boards, a corner stake, and plumb bob to establish building lines. A–Plan view. B–Lines must intersect over the nail in the corner stake.

Batter boards consist of three heavy posts and two horizontal boards. The posts hold the boards solid.

The following description refers to the numbers on the drawing. First, at number 1, the posts are driven into the ground. They are placed beyond where the hole will be dug.

A surveyor's level is used to measure the elevation. The elevation is marked on each corner post at number 2.

The batter boards are nailed to the center post. The top of the board is even with the elevation mark. Next, a carpenter's level is used to level the batter boards. Then, the ends are nailed to the outside posts.

A heavy string is stretched above the nail in the corner stakes. A plumb bob is used to line up the string and the nail. The batter board is marked where the string crosses it at number 4.

A nail or saw cut is put in the batter board at the mark. A string or wire is firmly attached.

Other points can be marked. The footing line at 5, and excavation lines at 6 may be added. The workers clearly label all marks.

After batter boards are marked on all corners, the strings or wires are taken down. Building can now begin.

PREPARING THE SITE

The engineering section designs the work area layout. They plan how to get in and out of the site. Temporary buildings for the workers are often placed on the site. They plan how to get utilities to the site. Lastly, the site is made secure.

Access to the Site

Most buildings are built close to streets and highways. Access to them does not pose much of a problem. Dams, pipelines, power plants, and offshore structures can cause access problems. Special transportation or roads may be needed. Workers on offshore projects ride helicopters and boats to work. Barges bring machines and materials to offshore jobs, Fig. 14-23.

Fig. 14-23. This is a floating work site. A bridge is being built. No fences are needed. Where are the offices and materials stored? Where do they get their utilities? (U.S. Department of Transportation, Federal Highway Administration)

Rerouting Traffic

When repairs to roadways are made, traffic must continue. Roadways may be *rerouted*. See Fig. 14-24. A temporary railroad bridge is called a "shoefly. Even streams are rerouted when dams are built.

Fig. 14-24. A large portion of the cost for this bridge was to reroute the road.
(Maine Department of Transportation)

Temporary Shelter

Projects that take a long time to build need *temporary buildings*. These buildings are used for many things. A temporary building can be an office, restroom, repair shop, or a laboratory. Laboratories are used to test soil and materials. A building may even be used as a temporary home. Temporary shelter is shown in Fig. 14-25. These buildings may be on wheels. A trailer is often used as an office. Trailers provide storage space.

A project may cover nearly all of the site. This causes special problems. Where are the office and storage spaces in Fig. 14-26? The street is sometimes used. Special permits are required. How is the street used in Fig. 14-26?

Getting Utilities

Building sites need electricity, water, and telephones. For sites in or near cities, this is not a problem. It is easy to get utilities to the site.

In remote place, Fig. 14-27, a diesel powered generator is used to provide electricity. Engineers plan how to get the power to where it is needed.

Water may be pumped from wells, lakes, or rivers. It must be *purified* (made clean) for drinking.

Fig. 14-25. Twelve workers live in this van. Each person has a bed, power, and bathroom with a shower. (Venture Ride Manufacturing, Inc.)

Fig. 14-26. Large buildings are sometimes built in "tight" places. Space for materials on street level is scarce. (The Stubbins Associates, Inc.)

Communication on the site and to distant places is needed. Telephones and radios are used.

Making the Site Secure and Safe

Work sites can be dangerous. They may attract vandals, thieves, and the curious observer. Something must be done to protect workers and the public, and reduce theft and distractions. The most secure sites are closed off from the outside. Solid walls are built on the site in Fig. 14-28. People can enter and leave the work site at certain places. Only workers and others bringing materials and equipment can enter. When work ends, the gates are closed and locked. People at the gate check to see who goes in and comes out.

Fig. 14-27. The utilities for this job site are all on board. They generate their own electricity. Communication is by radio. (Raymond International, Inc.)

Fig. 14-28. Barricades (fences) are built around construction sites.
(Earl R. Flansburgh and Associates, Inc.)

Safety for workers is planned in two ways. First, safe methods are practiced. Second, safety equipment is used.

Safe workers know how to work safely. They have the right tools and machines. Lastly, they *use* what they know and have. Training programs give workers the knowledge. The workers buy tools, or the firm provides the tools and machines. Workers are responsible for improving their own attitudes. Safe workers maintain good attitudes.

SUMMARY

A building permit is the first requirement when preparing to build. A set of plans is submitted to get the permit. Then site clearing can begin. Structures and trees can be salvaged or destroyed.

New structures are located using surveying methods that start with a centerline for highways or with a baseline for buildings. Stakes are set in the ground for cut and fill instructions, and reference points are placed on batter boards. Temporary shelter is built for workers and supervisors, and electrical power is set up.

Safety on the site is very important. Fences and walls help keep children out. Also, safe work methods are stressed.

KEY WORDS

All of the following words have been used in this chapter. Do you know their meaning?

Baseline
Batter board
Blasting
Breaking
Center baseline
Centerline
Convert
Cut and fill
Elevation
Landmarks
Moving
Offset baseline
Rerouted
Rubble
Salvage
Slope stake
Stop work notice
Tax assessor
Temporary building
Wrecking

TEST YOUR KNOWLEDGE

Write your answers on a separate sheet of paper. Do not write in this book.

1. Building standards required for a building permit are stated in the local _____ _____.
2. Getting a building permit for a new structure requires complete _____.
3. Brush can be removed with a cable or chain that drags an iron _____.
4. Land elevation stakes show the amounts of _____ and fill to be done.
5. One use of _____ stakes is to show the width of cut or fill for a road.
6. Reference point markers on posts set up on a building site are called _____ _____.
7. The location of any corner batter boards is on the (inside/outside) of a building corner.
8. Use a _____ _____ to line up a string being stretched above a building corner stake.
9. Living arrangements for workers are provided by _____ shelter.
10. You can help keep children away from hazards with walls and _____.

Matching questions: On a separate sheet of paper, match the definition in the left-hand column with the correct term in the right-hand column.

11. _____ Selling items saved from a site.
12. _____ A survey line used for a highway.
13. _____ A survey line used for a building.
14. _____ Removing a building by striking or bulldozing.
15. _____ A line from an existing survey.
16. _____ Removing a building with timed explosions.

a. Wrecking.
b. Baseline.
c. Boundary line.
d. Blasting.
e. Centerline.
f. Salvaging.

ACTIVITIES

1. Sketch out a clubhouse or storage shed. Lay it out at home or at school. Use the center axis or baseline.
2. Lay out a sidewalk to the clubhouse. Stake the centerline and the edges. What would you need to clear the site? How would you design the workplace?
3. What would you do to make a site like the clubhouse site secure? How long will it take? How much will it cost? Is it worth it?

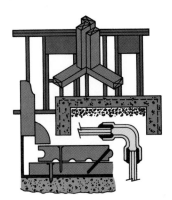

Chapter 15

DOING EARTHWORK

After studying this chapter, you will be able to:

☐ Explain how excavations are stabilized against cave-ins.

☐ Describe some soil removal methods both under water and on land.

☐ Define construction language like: scraper, motor grader, backhoe, and scarifying.

☐ Give reasons for choosing dredging or caissons when working under water.

☐ Give reasons for choosing trucks or conveyors when moving soil over rough terrain.

☐ Discuss selling or dumping of soil.

☐ Describe finish steps such as trimming and compacting.

Earth is made up of rock and soil or a mixture of the two. See Fig. 15-1. Its density (weight of 1 cubic yard) varies.

Bank soil is soil that has not been disturbed. It is quite dense because it has little air in it.

Loose soil has been disturbed. It has been broken up and contains air. It takes up more space with air in it. The amount it expands is called *swell.* Loose soil is not a good support for structures.

Compacted soil has been purposely compressed to drive out most of the air. It is more dense than bank soil. Heavy machines may be used to compact it. Fig. 15-2 compares the three states or conditions of soil.

EARTHWORK

To prepare a site for construction often means that some changes must be made in the earth. The shape or contour may be changed because it is too hilly or too flat. Poor soil may need replacing. Even after rearranging soil on a site, there may be too

A

B

C

Fig. 15-1. Three types of earth are common. A—Rock and soil mixture. (Caterpillar Inc.) B—Soil containing organic matter. C—Rock (Ingersoll-Rand Co.)

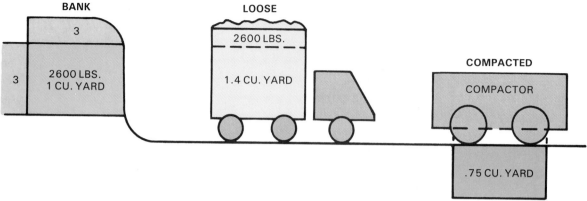

Fig. 15-2. Bank, loose, and compacted volume differ. Loose soil has more air space than bank or compacted soil.

much, and some must be removed. The changes made are called *earthwork*. Earthwork includes eight processes: stabilizing, loosening, excavating, adding, moving, disposing, compacting, and finishing.

Stabilizing Earth

Stabilizing is done to soil so that it will not cave in when soil around it is removed. Cave-ins can injure people and damage machines. A cave-in can cause extra work, too. After a cave-in, soil must be removed. Repairs to machines and the site may be needed.

Digging large holes can cause damage to buildings nearby. These buildings put pressure on the soil. This may cause the footings to crack and damage the buildings. The problem is the worst when nearby buildings have shallow footings. The footings can slide into the hole beside them.

Builders have ways to prevent cave-ins. They can put slope on the sides of the holes. *Retaining walls* are also used to support the soil around holes. Manufactured devices are used for small holes. The bottom is open so soil can be removed. The ends are left open so sewer pipes can be connected. When work is done, the box is moved to the next section. Larger holes require retaining walls with sheathing, wales, and bracing, Fig. 15-3.

Coffer dams, Fig. 15-4, are used to keep water out of holes. Steel sheathing is driven into the soil. The pieces of sheathing are joined to seal out most of the water. The water that seeps in is pumped out. The sheathing is held in place with bracing and shoring.

Loosening Soil

Hard earth is freed through *loosening.* Rock is *blasted,* Fig. 15-5. Frozen, hard, or rocky soil is *ripped.* See Fig. 15-6. Steel prongs are pulled through the soil. The prong can plow 3 ft. into the

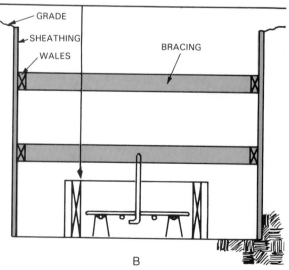

Fig. 15-3. Two methods of providing retaining walls.
A—A steel box is used for small holes. (Caterpillar Inc.) B—Larger excavations require the building of shoring.

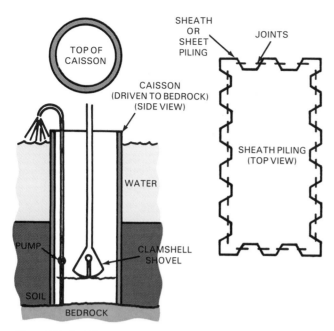

Fig. 15-4. There are many ways to make coffer dams. Two methods are shown in the drawing.

Fig. 15-6. A ripper loosens hard rocky soil. One or two prongs are forced into the ground and pulled forward. (Caterpillar Inc.)

soil. Scarifying is like ripping but does not go as deep. The purpose of *scarifying* is to loosen soil 1 to 6 in. (2.5-15 cm) deep.

Air hammers are used to break boulders. One is being used in Fig. 15-7.

Excavating

When you are *excavating,* you are removing earth from a space. Digging a hole for a basement is

an example. A special excavating method is needed for tight work areas. Most buildings in cities are only allowed space below the ground. Large open holes called pits are dug in the earth. Machines doing the digging and hauling are in the hole. This method is called *bulk pit excavating.* The soil taken out is called *spoil.* Spoil is loaded onto trucks and removed.

Backhoes, Fig. 15-8, can work from the edge of the hole. Basements of homes are dug this way.

Fig. 15-5. Holes are drilled and explosives are put into the holes. The blast loosens the soil. (Ingersoll-Rand Co.)

Fig. 15-7. This rock breaker is operated with air. Others use direct drive from engines. (Ingersoll-Rand Co.)

Fig. 15-8. Backhoes are useful machines. They can dig trenches or basements, break concrete, or load trees. (Caterpillar Inc.)

Spoil is piled up or hauled away. The piles of spoil are used later to finish the site.

Sometimes harbors and rivers need to be deepened. Suction dredges are most often used, but scoops do some jobs. Just as with the suction method, digging equipment is mounted on barges, Fig. 15-9. Equipment includes shovels and clamshell scoops. A *scow* (barge that unloads from its bottom) hauls away the spoil.

Fig. 15-9. Dredging is removing soil from under the water. The digging machines are mounted on barges. (American Dredging Co.)

Tunneling is a way to get useful space below ground. Subways are built this way. The parts of a subway are shown in Fig. 15-10.

Cut and cover is a method of building tunnels. A trench is dug. The floor, walls, and ceiling are put in place. Utilities are added. The outside of the tunnel is waterproofed, and then it is covered with soil. See Fig. 15-11.

Tunnels dug from below ground are bored. They start from the ends and from shafts. A shaft is a tunnel that goes straight down. Workers, machines, and materials get to the tunnel through the shafts. Soil and rock, called *muck,* is taken out of the shafts. One process begins at area A in Fig. 15-12.

Boring proceeds both ways from the shaft. Work goes on until the tunnels from two or more shafts are joined.

Rock is drilled and blasted. See area B in Fig. 15-12. Muck is loaded and hauled out. All soil is hauled out at area C in Fig. 15-12.

Steel or concrete liners are pushed out against the tunnel walls. The tunnel is now locked in the earth near B in Fig. 15-12.

When the tunnel goes below the water table (existing ground water level), special equipment and water pumps are used. The work space is kept under pressure to keep the water out. The workers can only work for short periods of time.

A common reason to remove soil is to get to a good *bearing surface.* Bank soil will serve as a bearing surface for low buildings. The footings need to be below the frost line. The *frost line* is the depth the soil freezes in wintertime. Backhoes are used to dig trenches and footings. See Fig. 15-13.

Bridge towers and large buildings require a stronger base. Bedrock is best. Soil many feet deep may cover the bedrock.

Caissons are driven into the surface of the earth. *Caissons* are hollow tubes of steel or concrete. Clamshell scoops on a crane are used to remove the spoil. The man in Fig. 15-14 is running a crane with a clamshell scoop.

Drilling makes cylindrical (tubular) holes. In Fig. 15-15, tapered steel piling is driven into predrilled holes. They are later cut to height and filled with concrete.

Utilities are laid in *trenches.* These long narrow holes are excavations, too. Backhoes are well suited to digging trenches. Fig. 15-16 shows a trench for a sewer line. The machine in Fig. 15-17 digs trenches quickly. The direction and depth of the digging are closely controlled.

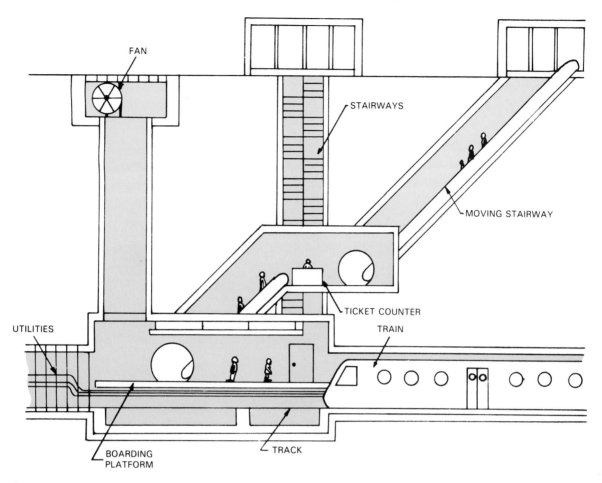

Fig. 15-10. A subway system has track tunnels, ticket counters, fan shafts, utilities, and stairways.

Fig. 15-11. Cut and cover tunnels are not deep. (ARMCO Construction Products)

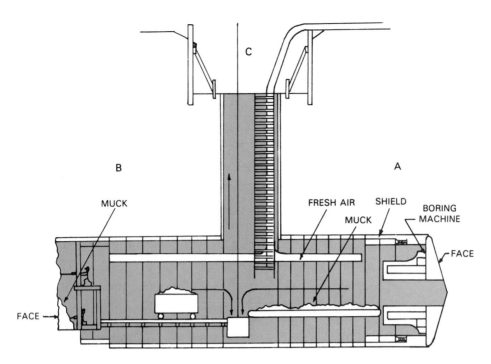

Fig. 15-12. Tunnels are bored from the ends or from shafts. The machine in area A is for boring into soil. Tunneling into rock is shown near B. The shaft is at area C.

Fig. 15-13. Backhoes are easy to maneuver. They are used to dig holes and trenches of all kinds. (Caterpillar Inc.)

Fig. 15-14. Clamshell scoops are used to dig holes in limited space. (Tennessee Valley Authority)

Fig. 15-15. This machine drills and drives tapered steel piling. (Raymond International, Inc.)

Fig. 15-16. A trench with a rounded bottom is being dug using a rounded scoop. Sewer lines will be placed in the trench. (Caterpillar Inc.)

Fig. 15-17. A trench for a pipeline is being dug with a rotating scoop wheel. The soil will be pushed back into the trench after the pipe is laid. (Barber-Greene Co.)

Trenches below water are dug by **dredging.** Soft soil is removed with a hydraulic dredge. Soil is loosened with a cutter, and high speed water removes the spoil. The water and spoil are pumped through pipes to land. The water runs away and spoil stays. See Fig. 15-18.

Fig. 15-18. Dredging. A round cutter loosens the soil. Spoil and water are mixed. The pump can move the spoil up to 15 miles. (Kenner Marine and Machine, Inc.)

Shallow excavations are made by pushing and scraping. Dozers are used to push soil. A reservoir can be formed this way. Soil from the center is pushed out to form the edges. The bottom is lowered and the edges are raised to form a dike. The reservoir in Fig. 15-19 was formed this way.

Fig. 15-19. This reservoir has a lot of water storage space. The liner will not let the water erode the banks. (Raymond International, Inc.)

Fig. 15-20. Scrapers are useful in excavating borrow pits or leveling. (Caterpillar Inc.)

Scrapers (commonly called earthmovers) can load, haul, and spread earth. Fig. 15-20 shows one in use. The parts and how they work are shown in Fig. 15-21.

Scrapers make cuts from 0-6 in. deep. The forward movement pushes the earth into the bowl. The apron is closed when the bowl is full. The cutting edge is raised and the load is hauled away. To unload, the apron is raised. The ejector pushes the load forward and the soil is spread. The thickness of spread is controlled by:

- How high the apron is raised.
- How fast the soil is expelled.
- The rate of travel.
- The height of the cutting edge.

Adding Earth to a Site

Earth is added to some sites. One reason is to raise the grade. A second is to get better soil on the site.

To raise the grade, soil is taken from one place and put on a second. When building roadways, the cut and fill method is used. Earth is cut from high spots to fill the low spots.

If there is not enough earth in the cuts, *borrow pits* are used. Soil is taken from one spot where it is not needed, and it is spread on the low spots. Borrow pits are often made into ponds for fishing and boating. See Fig. 15-22.

Storms can ruin beaches when the sand is washed away. Hydraulic dredges are used to restore them. Sand and water are pumped from offshore onto the beach. The water runs away and leaves the sand.

The land in a swamp makes a poor road. A channel can be dredged in marshland. Soil mixed with water is dredged from another place and is pumped to the site. The soil raises the roadbed above the

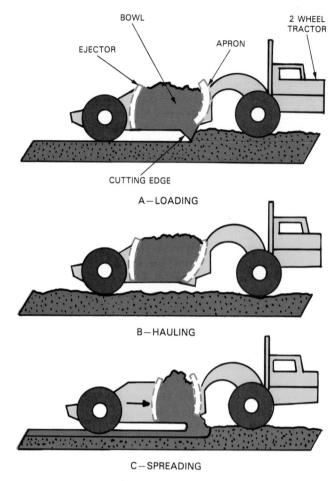

A—LOADING

B—HAULING

C—SPREADING

Fig. 15-21. Scrapers are large and look complex. Their operation is simple. Soil is pushed in, held, and pushed out.

Fig. 15-22. Soil was taken from the borrow pit to construct ramps on the new freeway. Muck (black area) was added to the pit to create a wetland. (Michigan Department of Transportation, photo by Tim Burke)

marsh. The stronger soil makes the road last. The expressway in Fig. 15-23 was built this way.

Dirt and gravel will pack and settle. Therefore, concrete and crushed rock are used under piping

Fig. 15-23. Dredging makes it possible to build roads in problem areas. Here cars travel where only small boats could go before. (American Dredging Co.)

Fig. 15-24. Train load after train load of soil was moved over rough terrain with the conveyor in the background. (California Dept. of Water Resources)

Fig. 15-25. This dredge is used to keep a cooling pond clear of soil. The pond is used to cool water from a power generator. The spoil is pumped hundreds of feet away. (Mud Cat Division)

systems. These materials will not let the pipes settle. Concrete holds a pipe rigid. It resists the force of water going around corners in the pipes. Without the bracing, pipes would crack. To make room for the crushed rock fill, extra soil is removed.

Moving Earth

Excavated spoil must be moved. One way is to use conveyors. *Conveyors* are moving belts or chains. They are suited to moving spoil over a short distance. A conveyor moved most of the soil for the earthen dam shown in the background in Fig. 15-24.

Fig. 15-25 shows a small dredge. The spoil is mixed with water and pumped to a dumping place.

When spoil is moved by trucks, cycle time is figured. One cycle includes loading, traveling, dumping, coming back, and maneuvering for loading.

Trucks are easy to load. They can move fast and dumping is quick. They are easy to get in and out of a place. All of this equipment reduces the cycle time. Fig. 15-26 shows some equipment.

The scraper shown in Fig. 15-20 does all tasks. It usually loads itself. Sometimes a bulldozer is used as a pusher.

Disposing of Earth

Earth is used, dumped, or sold. In doing cut and fill work it is used rather than dumped or sold. Riv-ers are dredged so larger ships can use them. The spoil can be used to raise a river bottom to make a preserve for wildlife. Land can be raised higher and used for road and factory sites.

During construction, topsoil is dumped on the site. Later, it is spread. Next, grass, trees, and shrubs are planted, Fig. 15-27.

Spoil is dumped in old pits and other low spots when it cannot be used. In time the dump site may be filled and made into useful land.

Fig. 15-26. Dump trucks come in a variety of sizes and designs. Even though they are large, they are easy to drive. (Caterpillar Inc.)

Some people need fill and will pay for the earth and the hauling. In this case, spoil is sold.

Compacting

Compacting takes the air out of the soil. Three compacting machines are shown in Fig. 15-29. Hand compactors are used for small jobs. They are good for places that are cramped. Some machines vibrate (shake). This helps to work the air out of the soil. Recall that compacted soil is more dense than most bank soil.

Finishing Earthwork

Washing, trimming, and grading are used to finish earthwork. Rock is washed to clean its surfaces. It is done by hand with high pressure water. Concrete bonds better to clean rock than to dirty rock.

A hole may be trimmed. To get the exact size and shape, work is done with hand tools. The holes in Fig. 15-28 were trimmed.

Grading is another finishing process. It smooths and levels soil. Grading is used to prepare a road bed. The slope in Fig. 15-30 is being graded with a machine called a motor grader. The driver follows

Fig. 15-27. A large concrete water tank is buried on this site. The soil was left at the site. (DYK Prestressed Tank, Inc.)

Fig. 15-28. This is compacted soil. It is strong enough to be used as forms for these footings. The holes were trimmed to exact size and shape.

B

A

C

Fig. 15-29. There are many machines used to compact soil. A–Vibrating plate.
B–Sheep's foot compactor. C–"Smoothwheel compactor." (Ingersoll-Rand Co.)

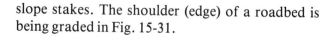

Fig. 15-30. Motor graders can form slopes or level sites. The blade can be set in many positions.

Fig. 15-31. A flat roadbed for a full depth asphalt median is being graded. The depth of cut is controlled by the wheel on the road. (CMI)

slope stakes. The shoulder (edge) of a roadbed is being graded in Fig. 15-31.

SUMMARY

Earth is rock, soil, or a mixture. Its density, swell, and compactibility can be measured. Earth is

stabilized so it will not cave in. Some earth must be loosened before it is excavated. Builders excavate to remove soil from a space. They also excavate because they need soil somewhere else.

Builders move the earth they remove. Builders have the option to dump, use, or sell the earth. The site is then finished.

KEY WORDS

All the following words have been used in this chapter. Do you know their meaning?

Backhoe
Bank soil
Borrow pit
Cassion
Coffer dam
Compacted soil
Compacting
Conveyor
Cut and cover
Dredging
Drilling
Earthwork
Excavating
Frost line
Loose soil
Loosening
Muck
Retaining wall
Scarifying
Scow
Scraper
Spoil
Stabilizing
Swell
Trench

TEST YOUR KNOWLEDGE

Write your answers on a separate sheet of paper. Do not write in this book.

1. Soil that has not been disturbed is called _____ soil.
   a. stabilized
   b. bank
   c. base
   d. cured

2. Sides of trenches in soil can be stabilized against cave-ins by giving them a _____.
3. Water is kept out of deep excavations with _____ dams.
4. What is the purpose of a caisson?
5. Soil under water is most often removed with a (clamshell scoop/dredge).
6. To get enough soil when needed, it can be taken from a _____ _____.
7. Crushed rock and also _____ are put under a pipe to keep it from settling.
8. A pusher sometimes follows a _____ to help it.
9. Compacting takes the _____ out of soil.
10. A slope or a roadbed is graded with a _____ grader.

ACTIVITIES

1. Describe the excavations needed for a sidewalk. What kind of tool would you use? How would you move the spoil and where would you dispose of it?
2. How would you finish the excavations? Would you need to add any earth material? Where would it come from?
3. How is it finished?
4. Find out the price you must pay for clean topsoil. The price is stated for a cube that is 1 yd. x 1 yd. x 1 yd. (A cube with these dimensions is one cubic yard.)

Chapter 16

BUILDING FOUNDATIONS

After studying this chapter, you will be able to:
☐ Explain the purpose of foundations and describe the five types.
☐ List requirements for footings.
☐ Explain how wood can be used in foundations.
☐ List parts of forms for concrete.
☐ Describe drain systems for footings.
☐ State the purpose of drainage systems, plastic sheets, and parging.

A structure consists of a substructure and a superstructure. The substructure is called a foundation. The foundation extends from the bearing surface to the main structure. The two parts may meet under the ground, at the grade level, or above ground. This concept is shown in Fig. 16-1.

The chapter covers the parts of foundations. Five kinds of foundations are described, and information on wood foundations is covered. You will learn how to make concrete and wood foundations.

A foundation supports the weight of a structure. Structures have to withstand the pressure of wind and water. Foundations work like the roots of trees. A stable foundation anchors the structure to the earth. See Fig. 16-2.

Two parts are common to all foundations. They are the bearing surface and the footings. These parts are shown in Fig. 16-2. The *bearing surface* is where the structure and the earth meet.

Small wooden structures are light in weight. Firm soil that has not been disturbed is a good bearing surface. Sandy or wet soils make poor bearing surfaces. They are too soft.

Large, heavy structures need a strong bearing surface. Bedrock or hard clay is best.

The *footing* is the bottom of the structure. Footings are designed to spread the weight of the structure over a greater area. In this way, a bearing surface can support more weight. It is the designer's job to match the footings with the bearing weight.

Most foundations have upright supports. *Upright supports* transfer the pressure of the structure down to the footings.

Walls and columns are upright supports, Fig. 16-3. Outside walls have to withstand three forces. The weight of the building pushes down. The soil pushes in. Water puts pressure on walls and soaks through them into basements.

The other upright support is the column. Columns are used to support the inner parts of the structure.

KINDS OF FOUNDATIONS

Most foundations can be classified as one of the following.
• Spread.
• Floating.
• Friction pile.
• Bearing pile.
• Pier.
The kind of foundation used depends on two things, the forces put on the foundation and its bearing surface.

Spread Foundations

Most small buildings use the *spread foundation.* See Fig. 16-4. Flat concrete footings are built under foundation walls and columns. The footings are wider than the walls. The weight of the building is spread over more surface. More soil is used to resist the pressure.

Concrete and pressure treated lumber are used for foundations. Pressure treated lumber does not rot. Insects will not bother it.

A

B

C

Fig. 16-1. Foundations can be below ground, at ground level, or above ground. A–Friction pile foundation. Tanks will be installed on concrete caps. (Raymond International, Inc.) B–Bridge foundations extend above the water. (American Institute of Steel Construction) C–A dam is nearly all a foundation. Buildings below the dam are superstructures. (Guy F. Atkinson Co., Ron Chamberlain)

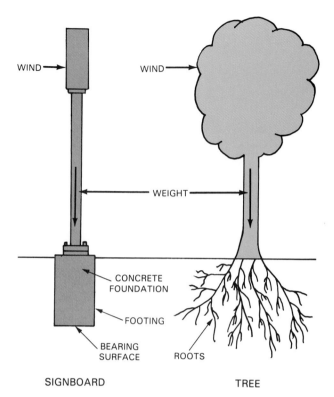

Fig. 16-2. Trees and signboards need a strong foundation. They must resist wind and support weight.

Fig. 16-3. In this structure, walls are used to transfer the weight to the footings. The webs (holes) in the concrete blocks are filled with concrete grout. Before the grout sets, rebar is placed in corner holes and every sixth hole to add strength.

The foundation wall is built on top of the footings. Poured concrete, or concrete blocks are most often used. Treated lumber and plywood work well also.

The footings are placed below the frost line. The frost line is the depth that soil freezes in the winter. When soil freezes, it expands. Soil expands with such force that it will move a structure. Over time, the footings can be broken, which will weaken the building. Soil does not freeze in the southern states. In northern states soil freezes more than four feet down.

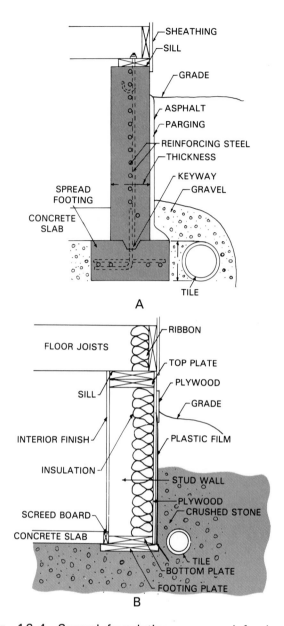

Fig. 16-4. Spread foundations are used for low structures. They require a firm bearing surface. A–Concrete. B–Crushed stone.

Floating Foundations

Light buildings built on weak soil use *floating foundations*. See Fig. 16-5. A single concrete slab supports the building. Steel rods or wire are used to strengthen the concrete. The slab needs extra strength under walls and columns. More thickness and steel are added at these points.

Friction Pile Foundations

Friction piles are used in weak soil. See Fig. 16-6. Friction piles never reach firm soil. Long poles are driven into the ground. They may be driven straight or at an angle.

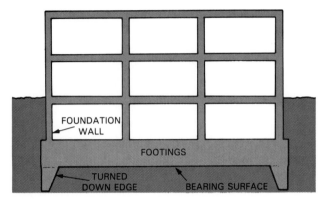

Fig. 16-5. Floating foundations are used on weak soil. The entire area under the building is used for support.

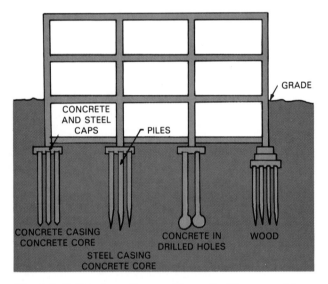

Fig. 16-6. Friction piles are like nails. They are driven into the material. Friction holds them in place.

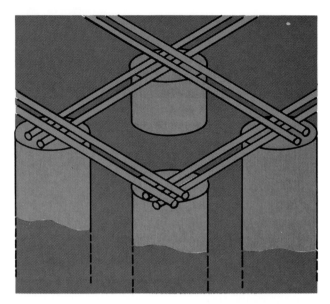

Fig. 16-7. A cap is used to spread the pressure over several friction piles. Concrete will be poured over the steel and the piles.

Fig. 16-8. Driving wood pilings for a temporary railroad bridge. Wood that is saturated (soaked) with water will not rot. It will rot when it is partly dry. Adding chemicals to the wood keeps it from rotting at any time. (Misener Marine Construction Co.)

Friction between the piles and the ground makes them stable. Friction can be increased by using more piles. Piling that is longer or thicker has more friction. A wavy surface on the piling helps, too.

Friction piles are grouped in rows under walls. A cluster of piles is used under columns. Concrete and steel are used to connect the tops. See Fig. 16-7. This *cap* spreads the weight evenly over all of the poles.

Friction tubes are made of concrete or steel. They are driven into the ground. They are cut to height, filled with concrete, and capped.

Some piles are made of wood, Fig. 16-8. Trees are cut and their bark removed. They are then pressure treated so they will not rot.

Bearing Pile Foundations

Bearing piles transmit the weight of the structure to the bearing surface. The bearing surface can be bedrock or firm clay, Fig. 16-9. Bearing piles are longer than friction piles. Some are over 200 ft. long. The long ones are made of concrete and steel. Some shorter ones are made of pressure treated wood.

Steel beams in the shape of an "H" can be used. They are driven to the bearing surface. Fig. 16-10 shows steel beams being driven. They are driven in sections. Sections are welded together as they are driven.

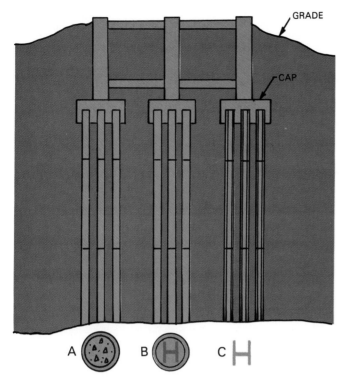

Fig. 16-9. Bearing piles are driven down to firm soil or bedrock. They are hollow pipes (A & B) or steel "H" piles (C). The hollow pipes are made of steel or concrete. They are filled with concrete after they are driven.

Fig. 16-10. A cluster of steel columns is being driven to bedrock. They will support a bridge. (Pennsylvania Department of Transportation)

Steel and concrete pipes are driven in lengths, too. They are later filled with concrete. Steel may be used to strengthen the concrete.

Bearing piles are used in clusters. A concrete and steel cap is used to spread the load.

Pier Foundations

Piers transmit the weight of the structure to a bearing surface. See Fig. 16-11. Many shapes are used. They are larger in diameter than most piles.

Piers are made by removing a column of earth. See Fig. 16-12. Drills are used for small round piers. Clamshell scoops are used for larger ones. The earth is replaced with concrete and steel. During excavation, steel or concrete liners are used. The liner keeps water out and keeps the sides from caving in. A single pier is used to replace a cluster of piling. Therefore, no cap is needed.

Fig. 16-11. The cable towers of this bridge sit on piers. They extend down to solid rock. (U.S. Department of Transportation, Federal Highway Administration)

Some piers do not reach a firm bearing surface. A special drill is used to remove the soil. It spreads out when it gets to the desired depth. See Fig. 16-12A. The bottom of the hole is bell-shaped. The larger base spreads out the load. It can, then, support more weight.

HOW SPREAD FOUNDATIONS ARE BUILT

This book is too short to describe all foundation types in detail. Only the spread foundation will be discussed further.

Spread foundations are made from concrete or pressure treated wood. In both cases, a trench is

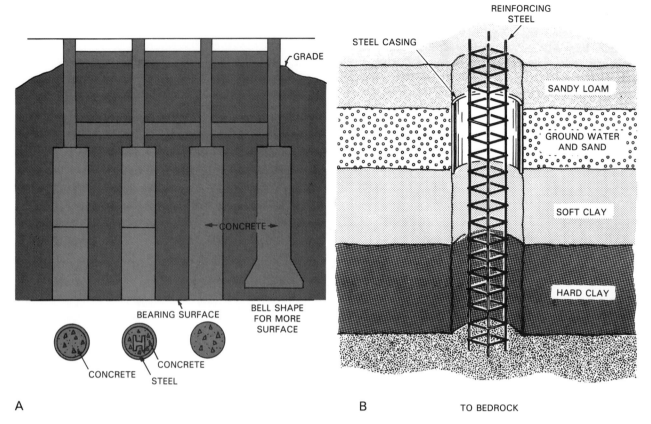

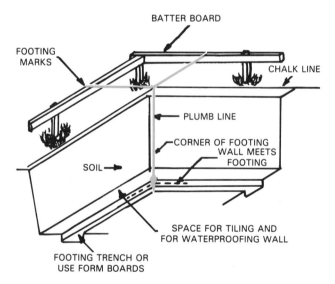

Fig. 16-12. Piers are used to support buildings or utility projects. A–Earth is removed to reach or to form a solid bearing surface. B–A caisson holds wet, loose soil in place until concrete is poured.

dug. The trench should reach below the frost line. If the structure will have a basement, a pit is dug.

A retaining wall is used if the soil is weak. What happens after the excavation is made depends on which material is used for the footing.

Building Concrete Foundations

As discussed before, a foundation consists of two parts–the footings and the foundation walls. The footings must be constructed prior to building the foundation walls.

Footings

If there is any water in the trench, it must first be pumped out. The bottom of the trench is trimmed. All loose soil is removed. Layout lines are restrung on the batter boards at the footing marks on the boards. A plumb bob dropped from the layout lines locates the footing forms. See Fig. 16-13. Recall that the same string system was used to place marks on the batter boards for later use.

The form for the footing is made of 2 in. (1 1/2 in.) lumber. The boards are placed on edge. There is a form board on the outside and inside. Footing forms are held in place with stakes driven into the

Fig. 16-13. The corner of the footing is found by dropping a plumb line from the building lines.

ground. The top edges of the forms are set level and at the exact height. Double-headed nails hold them in place. These nails are easy to pull when the forms are removed.

Reinforcing steel is placed next. The working drawing shows the size and location. Poured foundation walls require special steel rods. These rods are bent. They project above the form and become a part of the wall. A *keyway* (groove) is put down the center of the footing. See Fig. 16-14. A wooden strip with slanted sides is the keyway form. The keyway strengthens the joint between the footing and the wall. In other foundations, the top of the footing remains flat, as when concrete blocks are used for the wall.

Concrete is poured into the forms. Forms are removed the next day. Concrete is kept moist and above freezing for the first week. A plastic cover and straw works well.

Foundation walls

Foundation walls are usually made of poured concrete or concrete blocks. Poured walls need forms. Forms can be built on the site. They consist of plates, studs, walers, ties, braces, stakes, and facing. These parts are shown in Fig. 16-14. Manufactured forms come in pieces and are assembled on the site. Manufactured forms are used in Fig. 16-15.

The first step in setting forms is to lay them out. Again, the lines on the batter boards locate the corners of the building.

The outside wall forms are built. Steel rods are placed and tied with wire. Four tying methods are shown in Fig. 16-16.

Form ties come in many lengths and designs. One of the many kinds is shown in Fig. 16-17. They are used as spacers between the inside and outside forms.

After the steel rods are in place, the inside forms are built. They are set and held in place. Forms are made plumb and straight. A spirit level is used to plumb the forms, Fig. 16-18. A tight line is stretched along a wall to check straightness. The wall is aligned with the line. Braces and shoring hold the forms straight and plumb against the weight of the concrete.

Concrete is poured to the desired height and left to harden. In a day or so the forms are removed.

Concrete block walls use mortar to hold the block in place. Mortar is special concrete made with cement, sand, water, and lime. The mortar lets the bricklayer adjust the blocks. The blocks should be plumb and level. When the mortar hardens, the walls become solid. See Fig. 16-19.

Building Wood Foundations

Wood foundations are being used for small structures. They are made of pressure treated plywood and lumber.

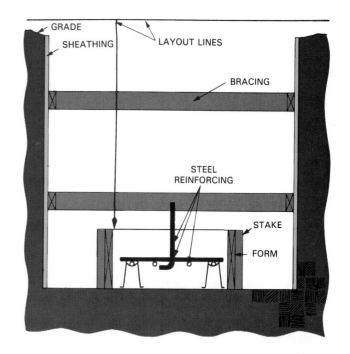

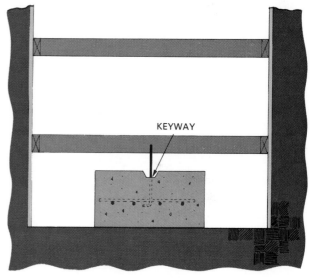

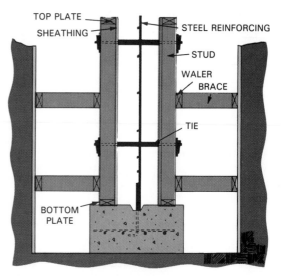

Fig. 16-14. Forms have plates, studs, walers, ties, braces, stakes, and sheathing.

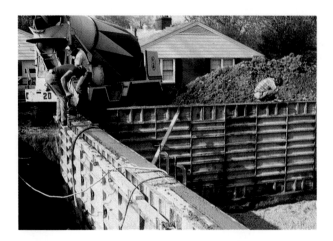

Fig. 16-15. Concrete is being placed in manufactured forms. The bracing near the workers holds the forms straight. (Mary Robey)

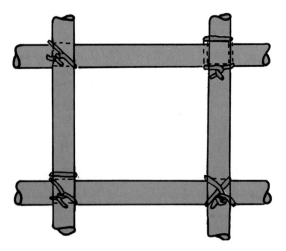

Fig. 16-16. Steel reinforcing bars must be held in place while other work is done. The method for tying them differs.

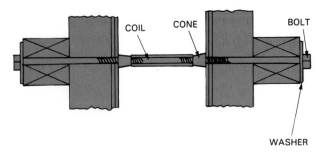

Fig. 16-17. Concrete forms are held apart with form ties. After the forms are removed, the ends are unscrewed to remove them. The holes in the concrete are filled.

Details for a wood foundation were shown in Fig. 16-4B. Crushed rock is used under the foundation. It lets the water drain away. The footing is a wide board or is concrete. The foundation wall is

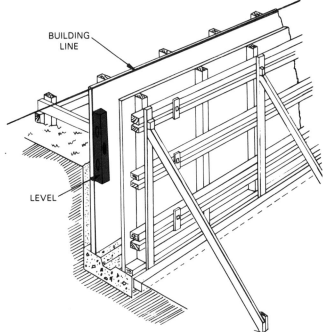

Fig. 16-18. A spirit level is used to plumb the forms. A tight line is used to straighten them.

Fig. 16-19. Concrete block foundation walls go up fast.

made of pressure treated lumber and plywood. Framing techniques described in the next chapter are used to build them.

WATERPROOFING FOUNDATION WALLS

Foundation walls must resist a third stress: water pressure. Water can soak through a solid concrete or wood wall. A foundation wall in contact with moist or water-laden soil can get damp inside.

Three methods are used to prevent damp walls–drainage systems, parging, and plastic sheets. *Drainage systems* remove the water from the soil around the building. Clay tile and plastic tubing are used. Water runs out of the soil into the tile or tube and into drains. Clay tiles are placed end to end. Water gets into clay tiles through small spaces between them. Water enters the plastic tubes through small holes punched in the sides. The drainage system is shown in Figs. 16-4B and 16-20. The system drains into the sewer.

Parging is put on the outside of concrete block walls, Fig. 16-20. Parging is a layer of mortar. This layer is covered with a coat of asphalt. *Asphalt* is a sticky, black oil product. Asphalt is troweled or sprayed on to the wall. Sometimes fibers are mixed with it. The fibers keep it from cracking.

Sometimes *plastic sheets* are used to waterproof foundations. Large sheets of plastic are hung on the outside wall. It keeps the soil from touching the wall. Fig. 16-21 shows a wood foundation. After it is erected, the outside wall is covered with plastic.

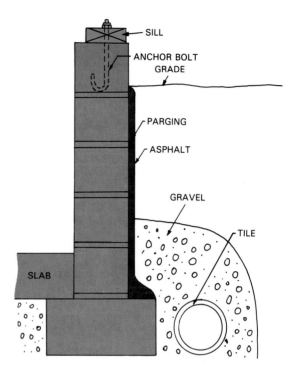

Fig. 16-20. Concrete block walls are waterproofed with parging and asphalt.

Fig. 16-21. Pressure treated wood foundations are built much like other wood frame walls. Tough plastic film is used to waterproof them. (Hickson Corp.)

SUMMARY

The foundation supports and anchors the structure. The kind of foundation used depends on the bearing surface and the forces of the structure. A bearing surface, footings, and upright supports make up a foundation. There are five kinds of footings. Concrete, steel, and wood are used in foundations. Drainage systems, parging, and plastic sheets are used to waterproof basement walls.

KEY WORDS

All the following words have been used in this chapter. Do you know their meaning?

Asphalt
Bearing pile
Bearing surface
Cap
Drainage system
Floating foundation
Footing
Friction pile
Keyway
Parging
Pier
Plastic sheet
Spread foundation
Upright support

TEST YOUR KNOWLEDGE

Write your answers on a separate sheet of paper. Do not write in this book.
1. The best earth types for supporting heavy structures are _____ and hard clay.

Matching questions: On a separate sheet of paper, match the definition in the left-hand column with the correct term in the right-hand column.

2. _____ Has flat footings wider than the walls.
3. _____ Transmits the weight to bedrock.
4. _____ Uses single concrete slab.
5. _____ Uses either bedrock or bell-shaped footing in clay.
6. _____ Uses pilings that never reach firm soil.

   a. Friction piles.
   b. Pier foundation.
   c. Spread foundation.
   d. Bearing pile foundation.
   e. Floating foundation.

7. It is best if forms for concrete are put together using _____-_____ nails.
8. What helps strengthen the joint between a footing and a poured foundation wall?
9. Use _____ _____ under a wood foundation to let water drain away.
10. The first coating put on the outside of a concrete foundation wall to seal against water is called _____.

ACTIVITIES

1. Ask an architect what kind of foundations are used under large structures in your town.
2. Find out where the frost line is in your area.
3. Design a tower that is 7 ft. above ground. What kind of foundation will you use?
4. Design a foundation for a clubhouse. Will it be concrete or treated woods.

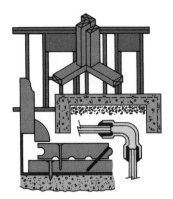

Chapter 17

BUILDING SUPERSTRUCTURES

After studying this chapter, you will be able to:
☐ Discuss highway projects, building walls, and wood and steel framing.
☐ Describe concrete paving machines, cut rock, and types of asphalt.
☐ Define studs, joists, plates, rafters, decking, and roof trusses.
☐ Sketch stud layout around a window or door.
☐ Define concrete and steel terms like precast, built-up, and tag line.

There are four kinds of superstructures. These structures are: mass, bearing wall, framed, and fabric structures.

Mass structures are solid or nearly so. They have little or no open space inside them. Dams are mass structures. *Bearing wall structures* have usable space in them. The walls hold the floors and roof above them. *Framed structures* have a skeleton. They are made of wood, reinforced concrete, or steel. A skin is used to enclose framed buildings. *Fabric structures* use air and cables to support specially treated cloth. Fig. 17-1 shows three kinds of members. Can you identify them? This chapter describes each structure.

BUILDING MASS STRUCTURES

Mass structures use large amounts of material. Most often, materials are in their natural state. Soil, rock, reinforced concrete, and asphalt paving are used. Soil is dug, hauled, spread/mixed, and compacted. Processing is not costly. The cost seems high because huge amounts are used. The project in Fig. 17-2 used millions of tons of soil. Cement was mixed with it to make it pack harder. It is called soil cement.

Rock is blasted or cut from a quarry. Blasted rock is broken into useful sizes and hauled to the building site for use. Rock is placed on shorelines so

Fig. 17-1. This building has mass, load-bearing, and frame members. Can you find them? (Gang-Nail Systems, Inc.)

Fig. 17-2. Soil cement was used on this project. Soil, cement, and water are mixed together. Cement makes the soil harder.
(Combustion Engineering, Inc.)

that the shore will not erode. It is used to build jetties and breakwaters. These structures control water flow and protect the shore from waves. Old concrete is recycled for the same purpose.

Cut rock costs more. It is only found in certain places. It is purchased from a supplier. Because it takes a lot of work to produce cut rock, the cost is high.

Reinforced concrete is made of cement powder, aggregate, water, and steel. Sometimes old concrete is crushed and used as aggregate. The retaining wall in Fig. 17-3 is made of concrete and steel. The steel rods are placed in the top and bottom. The concrete enters the form from the top.

When a concrete street is being placed, steel reinforcing rods are set. They are held off the subgrade with *chairs.* The paving machine spreads, screeds, compacts, and does some finish work on the concrete. Most finish work is done by hand.

The concrete is covered to keep it moist. If it dries too fast, the surface is weak. Burlap can be used to

hold moisture. Other builders spray on a curing compound. You can see this being done in Fig. 17-4. A *paving train* is shown in Fig. 17-4. The machines run on two tracks. There is one at the top and one at the bottom of the canal.

Asphalt paving is made of aggregate and asphalt. Aggregate is a mixture of sand and gravel. The asphalt holds the aggregate together. Coarse aggregate is used in the first, thick layer, Fig. 17-5. The finish layer uses finer aggregate, Fig. 17-6. Steel is not needed to strengthen asphalt paving. The paving is flexible and tough.

Fig. 17-4. A paving train has two or more machines. This one places and finishes concrete. The third machine is a walkway. The concrete is troweled and curing compound is applied from the walkway. (CMI)

Fig. 17-3. This retaining wall is a mass structure. Concrete goes in the top of the form.
(Miller Formless, Inc.)

Fig. 17-5. The first layer of an asphalt highway is thick. It will be tamped down with a compactor.
(Ingersoll-Rand Co.)

Fig. 17-6. The finish coat of asphalt is being placed on a parking lot.
(Puckett Brothers Manufacturing Co., Inc.)

Hot-mix asphalt hardens when it cools. The cold mix hardens when a solvent dries out of it. Both kinds are compacted after they are spread. Compacting squeezes air out.

BUILDING BEARING WALLS

Bearing wall structures have strong walls, Fig. 17-7. Older structures used rocks and blocks held together with mortar. Walls were 3 to 4 ft. thick and windows were small. *Arches* and *lintels* were used to bridge the openings.

Masonry walls are made of small units. They are held in place with mortar. The most common units are brick, block, and rock. See Fig. 17-8.

Fig. 17-8. Bricks and blocks were used to build this bearing wall structure. (Pella/Rolscreen Co.)

Load-bearing walls are seldom used in buildings with more than one story. It takes a lot of material and labor to build them. Masonry walls do not insulate very well.

Concrete block walls are used in industrial and commercial buildings. The materials cost little and walls go up fast. Like other masonry, they do not provide good insulation. It costs a lot to heat buildings made from block unless insulation is added to the inside of these walls.

Tilt-up bearing walls are made of reinforced concrete. The concrete floor for a building is cast. Forms for the walls are built on the floor, Fig. 17-9.

Fig. 17-7. This old church has walls 3 ft. thick. The structure has little space inside for the amount of materials used.

Fig. 17-9. Tilt-up bearing walls are made on the floor slab. Forms are built and steel mesh is set. Concrete is being placed on this site. (The Burke Co.)

Steel is set in the forms. Concrete is poured, finished, and cured. The forms are removed. Fig. 17-10 shows a wall section being lifted into place. Steel in the walls projects out the edges of the sections. Steel rods from two wall sections are welded together. Concrete columns are poured where the wall sections meet.

A

Fig. 17-10. Cast walls are lifted and placed with cranes. (The Burke Co.)

B

BUILDING FRAME STRUCTURES

Frame structures are made from wood, reinforced concrete, and steel. The frame in Fig. 17-11 is wood.

Frame structures are used in many ways. You can see several uses in Fig. 17-12.

Most of today's buildings are frame structures. The frame is made up of beams, rods, poles, or

C

Fig. 17-11. A wood frame was used in this structure. Wood is strong and light in weight.

Fig. 17-12. Many frames are not enclosed. These were built to support something. A—Most equipment is exposed to the weather. B—Tubular shapes will support themselves. (Southern California Edison Co.) C—A triangle shape is strong. (Bob Dale)

boards. The frame supports the roof and all floors, ceilings, and contents. Frame structures have many good features. Frame structures provide more space inside than with mass structure walls. They are light in weight. Frame structures use less materials. They are easy to insulate.

Small one to four-story buildings can use wood frames, Fig. 17-13. Taller buildings can use reinforced concrete frames, Fig. 17-14. The very tall structures use steel frames, Fig. 17-15.

Fig. 17-13. Wood frames are easy to make with simple tools. Roof trusses do not need interior bearing walls.

Fig. 17-14. Concrete and steel were used to make this office building's frame. Other materials were added during finishing.
(The Manitowoc Co., Inc.)

Fig. 17-15. Steel frames are used for the tallest buildings. (HNTB Engineers)

Wood Frames

Wood frames for buildings are built in four parts. These are floor, wall, ceiling, and roof frames. Frame members are placed 12, 16, or 24 in. on center. That way, 4 x 8 ft. sheets of plywood will fit without cutting.

Framing the floor

Many small buildings are built on a concrete floor. Houses built over a crawl space or basement use a framed floor. The method most often used is *platform construction.* The platform is shown in Fig. 17-16. The first part of the platform frame is the sill. The sill lies on top of the foundation. It is bolted to a concrete or block foundation. Insulation is placed between the foundation and sill, Fig. 17-17. It prevents air and insects from getting into or out of the building.

Other frame members are used to give strength. The members consist of 2 in. lumber 8 to 12 in. wide. The members are set on edge so they have greater strength to carry loads. *Joists* are fastened to the *joist headers* to keep them vertical. On long

A

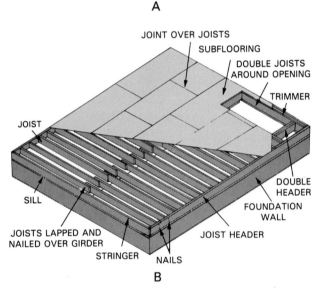

B

Fig. 17-16. Examples of platforms. A—Workers can stand on the platform. Wall framing is nailed to the platform. (J-DECK Building Systems) B—Detailed drawing shows parts of the platform and construction materials.

Fig. 17-17. Insulation is being placed on top of the foundation wall. A plank called a sill is held in place with the bolt. (CertainTeed Corp.)

spans they are spliced over a supporting girder or beam. The outside joists which rest on the wall are called *headers* and *stringers.* The platform provides a nailing base for the subfloor. The subfloor provides a base for the finished floor.

Wall framing members

Wall frames consist of studs, plates, headers, corner posts, and sheathing, Fig 17-18.

The *studs* are the upright supports. Spaces between the studs in outside walls are filled with insulation. Surface coverings are nailed to the edges of the studs.

Extra studs are needed at the corners and where walls meet. The extra studs are used for support and as a nailing surface. Both outside and inside materials are nailed at the corner. Corner and intersection posts are shown in Fig. 17-19.

New framing methods leave more space for insulation. The methods save on heating and cooling costs. Framing methods are shown in Fig. 17-20.

Plates are used as a nailing surface for the studs at the top and bottom. Double top plates add strength and keep walls straight. They also support the ceiling and roof.

Rough openings are made in walls for doors and windows. They are about 1 in. larger than the door and window units. Doors or windows are placed in rough openings. They are leveled with shims and fastened in place.

Sheathing is the covering for the structure. Sheathing is nailed to the outside of the wall. Plywood sheathing makes a building stronger. Fiberboard sheathing is not as strong but it insulates better. Foam plastic sheathing insulates well but is weak, Fig. 17-21. *Diagonal* (angled) *braces* are used

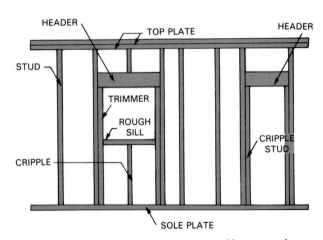

Fig. 17-18. Wall frames have top and bottom plates. Between them are studs, headers, cripples, trimmers, and rough sills.

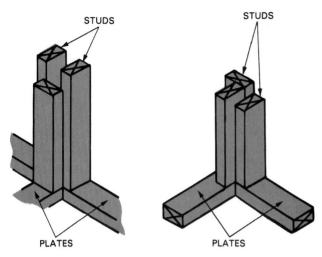

Fig. 17-19. Extra studs are used where walls meet. They add strength and provide a nailing surface for wall covering. Left. Corner framing at interior wall. Right. Corner for outside walls.

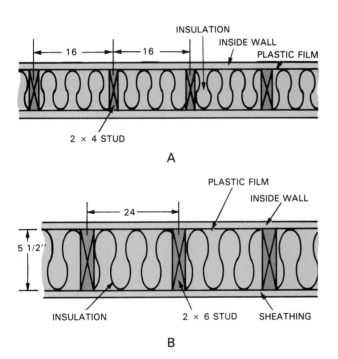

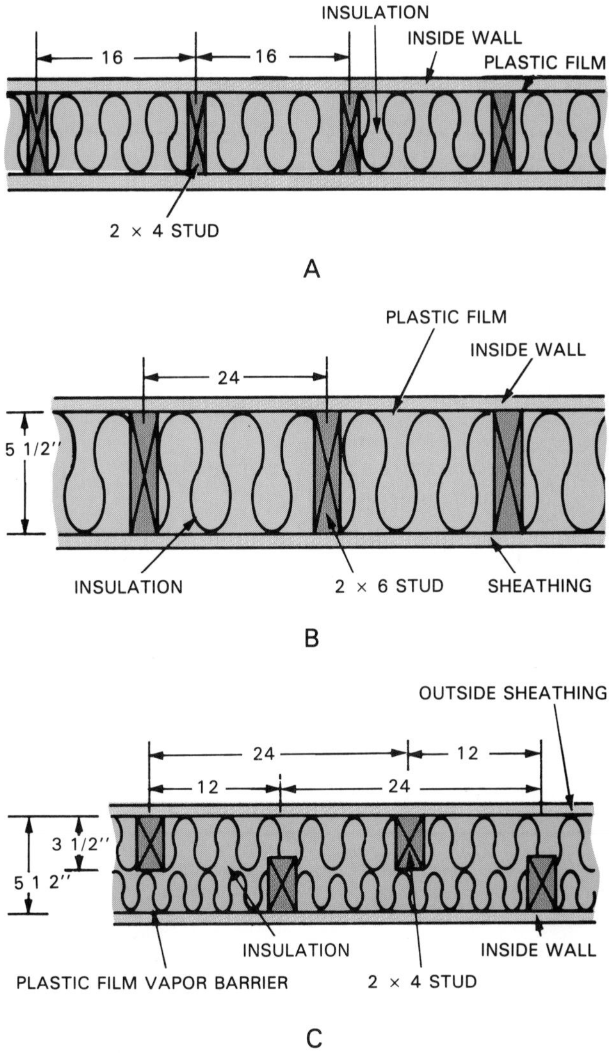

A

B

C

Fig. 17-20. The extra materials in a wall cost extra money. The savings in heating costs will pay for the materials. It may take several years. A–Standard wall. B–2 x 6 studs. C–Staggered 2 x 4 studs.

Fig. 17-21. The roof is sheathed with plywood. The walls are sheathed with foam plastic. The plywood is strong. The foam plastic insulates very well. (Sellick Equipment, Ltd.)

to strengthen the wall. Both wood and metal strips are used.

Framing ceilings

Ceiling framing ties the walls together. It supports the floor above and the ceiling below. One end of the *ceiling joists* rests on the outside walls. See Fig. 17-22. The other ends rest on the inside bearing wall.

Ceiling joists are much like floor joists. In fact, in a two-story building, a second story floor joist supports the ceiling for the lower floor. Ceiling joists serve as roof joists on structures with flat roofs. In a one story building they support less weight. Therefore, they are smaller and may be farther apart.

Framing roofs

Roofs cover the tops of buildings. Roofs are flat or pitched, Fig. 17-23. Flat roofs use roof joists covered with plywood. The plywood is called decking. Roofing methods are described in the next chapter.

Pitched roofs are designed in many ways. Parts of the roof framing are shown in Fig. 17-24. *Rafters* are the main framing member. The *ridge board* keeps them spaced right at the roof peak. A *collar beam* keeps them from spreading. Refer back to Fig. 17-22 for an example of a collar beam.

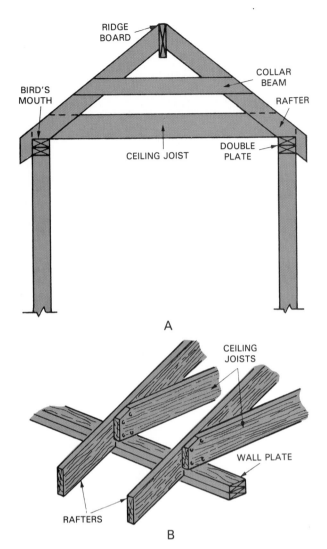

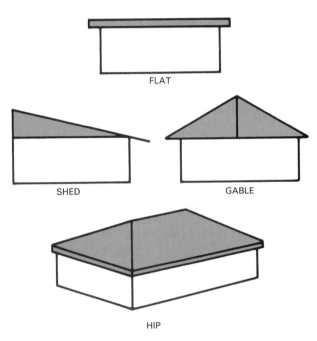

Roof trusses are used for roof framing. A roof truss has both rafters and ceiling joists. Roof trusses are shown in Figs. 17-1 and 17-13. Truss manufacturers fasten joints with steel plates, Fig. 17-25, or some other kind of special fastener to keep joints very rigid.

Reinforced Concrete Frames

Concrete is a good material for slabs and footings. It is also used for building frames. Concrete building frames can be cast-in-place or precast.

Fig. 17-22. What a joist is. A–Note how joist keeps walls from spreading outward. B–The top corners of ceiling joists are cut off. They are cut at the same angle as the rafter.

Fig. 17-23. There are many roof designs. Flat, gable, hip, and shed are four of the most common.

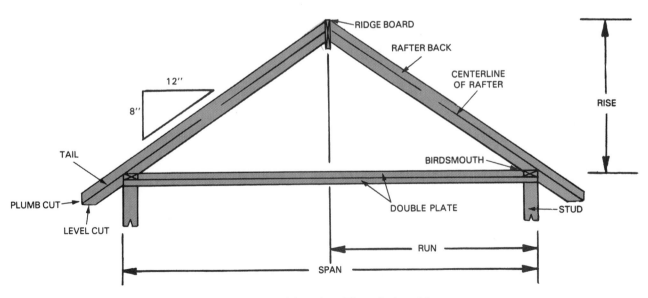

Fig. 17-24. The rafter is the main member in roof framing. The relationship of rise to span determines the pitch. This one is one-third pitch.

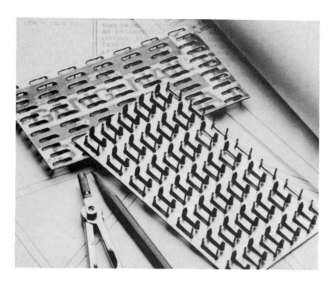

Fig. 17-25. Steel plates reinforce the joints of roof trusses. They are pressed or rolled into the trusses. (Gang-Nail Systems, Inc.)

Cast-in-place concrete

All concrete is placed and finished on the site in cast-in-place frames. The parking structure in Fig. 17-26 has a cast-in-place frame. As construction progresses, you can see jobs in all stages of completion.

Forms can be purchased. Others are built by carpenters. Wood forming methods are being used in Fig. 17-27.

Fig. 17-26. This parking structure is being built in stages. From the top down, you can see the shoring (A), reinforcing steel for walls (B), wall forms (C), and finished wall (D). (Ball Memorial Hospital)

Precast concrete

Precast concrete is cast from a form. Then it is placed in a structure. These members are cast in a

Fig. 17-27. Forms are built on the site by carpenters. (United Brotherhood of Carpenters and Joiners of America)

plant. They cost little and are high quality. Forms are used over and over. The process of precasting concrete is closely controlled. Some standard shapes are shown in Fig. 17-28.

At the site, precast parts are joined with cast-in-place methods. The building in Fig. 17-29 was built this way. Precast parts are used for the floors. The outside facing was precast, too.

Steel Frames

Steel frames make it possible to build skyscrapers and large bridges. They are also used in smaller buildings. The steel frame provides a high strength-to-weight ratio.

Steel structural members serve a second purpose in smaller buildings. The building is more fire resistant. That means that the building will not burn as readily. Steel framing was used for the structure shown in Fig. 17-30.

Steel framing parts are rolled steel sections or built-up members. Standard rolled steel sections

are in the form of letters. The letters I, U, H, and L are most common.

Built-up members have many shapes. The most common are H and I, squares, and rectangles.

Fig. 17-29. Using precast parts speeds up construction. Floor panels and walls were precast. (The Manitowoc Co., Inc.)

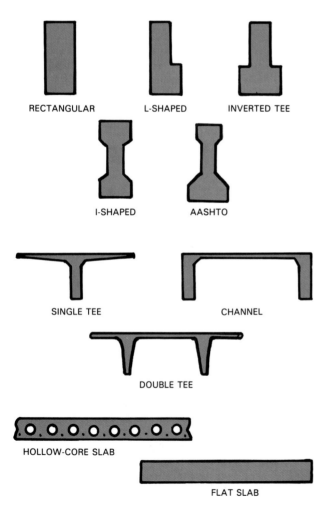

RECTANGULAR   L-SHAPED   INVERTED TEE

I-SHAPED   AASHTO

SINGLE TEE   CHANNEL

DOUBLE TEE

HOLLOW-CORE SLAB

FLAT SLAB

Fig. 17-28. Precast concrete parts are made in plants. There are standard shapes for beams and slabs.

Fig. 17-30. Steel framing and brick exterior make a building more fireproof. (Spectra-Physics)

Fig. 17-31. Over 100 bolts will be used to fasten the two ends of this bridge beam. (American Institute of Steel Construction)

Joists for houses can be I-shaped or they can be hollow rectangles.

Steel is joined by welding and bolting. The beam in Fig. 17-31 will be bolted in place. Both welding and bolting were used in Fig. 17-32. Girders and bracing are being joined to a column.

Columns are framing units that extend from footings and through one or more stories. Columns are bolted to the foundation. The anchor bolts are placed deep into the concrete footing. The bolts are used to reduce side movement and to plumb and set the height of columns. Space under the flange is filled with grout. See Fig. 17-33.

Cranes are used to lift steel parts into place. The operators must be highly skilled. Fig. 17-34 shows cranes in use. Placing the highest or last piece of steel in a structure is called "topping out." See Fig. 17-35. A ceremony sometimes follows.

Fig. 17-32. The joints in steel structures have to be very good. They must hold for many years. (Mitchell/Giurgola Architects)

Fig. 17-33. The bolts are used to plumb and set the height of the column. After they are set, grout is forced under the column. The grout maintains the column setting. (The Burke Co.)

Fig. 17-34. Cranes lift steel into place. The operators are highly skilled. (The Manitowoc Co., Inc.)

Erecting steel frames is risky. Gloves and hardhats are standard. Eye protection is worn by welders, cutters, and chippers. A chipper uses a hammer to clean a weld. People working above ground are protected from falls. Lifelines, safety belts, and nets are used.

Weather is a hazard. Wind can blow workers off balance. Steel beams can start to swing in the wind. *Tag lines* are used to keep beams from swinging. Look again at Fig. 17-34.

Rain and snow make metal parts slick. Oil and other hazards are removed before hoisting.

Fabric Structures

Using fabric for a building material is not new. It has been used for thousands of years in tents. Build-

Fig. 17-35. Putting up the final story of a building. Concrete will cover the steel work. (Associated General Contractors of America)

Fig. 17-36. Grain is stored in this air supported fabric structure. They are easy to build and cost less than concrete, metal, and wood buildings.

ings, in which fabric was a part, were first used as small temporary buildings. Now they are used for large sports arenas, zoos, and art centers.

Fabric is a building material in air-supported and tensile structures. Structures are made stable by inflating or removing air, or with masts and tightly stretched cables. *Air-supported structures* look like bubbles. An example is shown in Fig 17-36. With a slight vacuum (air is removed), the structure looks more like a tent.

Tall, strong masts and heavy cables form the frame of *tensile structures.* In these structures, fabric is stretched and clamped to the cables. These methods were used in the outdoor concert space shown in Fig. 17-37.

Fig. 17-37. A sheltered space and grand effect are created with this tensile fabric structure. (BIRDAIR, Inc.)

SUMMARY

Superstructures are usually above ground. They vary greatly. Mass structures have little space inside them. Bearing wall structures have strong walls. Cast concrete and masonry units are used. Most structures today use frames. A frame is the skeleton. Frame structures are made of wood, reinforced concrete, and steel.

KEY WORDS

All the following words have been used in this chapter. Do you know their meaning?

Air-supported structure
Asphalt paving
Bearing wall structure
Ceiling joist
Chair
Collar beam
Column
Diagonal brace
Fabric structure
Framed structure
Header
Joist
Joist header
Mass structure
Paving train
Plate
Platform construction
Rafter
Ridge board
Roof truss
Rough opening
Sheathing
Stringer
Stud
Tag line
Tensile structure
Tilt-up bearing wall

TEST YOUR KNOWLEDGE

Write your answers on a separate sheet of paper. Do not write in this book.

1. What is soil cement?
2. Steel reinforcing rods for concrete are held off of the subgrade with _____.
   a. lintels.
   b. masonry.
   c. tag lines.
   d. chairs.
3. Name the two kinds of asphalt.
4. Arches and _____ make a bridge over openings in bearing walls.
5. Masonry walls do not _____ very well.
6. Insulation is put between the foundation and the _____ of a wood frame wall.
7. Vertical wood members in a framed wall are called _____.
   a. studs.
   b. joists.
   c. headers.
   d. plates.
8. A short stud below a header for a door is called a _____ stud.
9. A ceiling joist is (vertical/horizontal/sloped).
10. Roof sheathing is attached to (rafters/joists).
11. Roof trusses need interior bearing walls. True or false?
12. A floor can be made with prefabricated or _____ concrete parts.
13. Steel has a higher strength-to-weight _____ than concrete parts.
14. A building using steel framing is _____ resistant.
15. Steel is assembled by bolting or _____.
16. For safety against falls, ironworkers can use _____, safety belts, and nets.

ACTIVITIES

1. Look in the basement or attic of your house. Find the parts of the frame. Can you name the parts? Use a camera or video camera to record what you find. Use the pictures to help explain framing parts to the class.
2. Study your home. What kind of frame was used to build it?
3. With a friend. Visit a store. Get permission from the manager. Record the kind of framing used in the structure. Consider using a camera or video camera. Use the pictures to help explain the framing system to the class.
4. Make a model of a small section of a wall frame. Use balsa wood for studs.

The steel frame for this structure is strong and takes up little room.

Chapter 18

ENCLOSING THE STRUCTURE

After studying this chapter, you will be able to:
☐ List types of roof coverings.
☐ Discuss how the roof angle helps to clear snow and rain.
☐ Discuss whether the cost of flat or angled roofs is the only way to choose a roof type.
☐ List parts of doors and windows and their frame openings.
☐ Describe how doors and windows are installed.
☐ Describe exterior wall coverings for insulation and decoration.

Structures are enclosed to keep out the wind, rain, cold, sun, heat, and dust. Enclosing the frame makes the contents safe from theft as well.

Some frames are not enclosed. They are designed to withstand the weather. The materials used do not corrode, or those that do corrode have their surfaces coated. The frame in Fig. 18-1 is painted, although rust would not disturb its use.

Builders enclose structures as soon as they can. Then, rain will not stop work. Damage is avoided if wood frames are kept dry.

Both roofs and sides of buildings must be enclosed. Roofs on low buildings are enclosed first. Windows and doors are installed next. Siding is applied last.

The lower floors of highrise buildings are enclosed before the frame is complete, Fig. 18-2. Steelworkers finish the frame one floor at a time. They are followed by concrete workers who cast the floor. Other trades install utilities and furnishings.

ENCLOSING ROOFS

Roof types may be flat, pitched, or curved. See Fig. 18-3.

Fig. 18-1. Frames are not always enclosed. Some frames are not for shelter. They only support something. (Exxon Corp.)

Fig. 18-2. The steel frames for the two towers are still being erected. Lower floors are enclosed. (Council on Tall Buildings and Urban Habitat)

A

B

Fig. 18-3. Kinds of roof. A—Curved roof. B—Flat Roof. (Pella/Rolscreen Co.)

Fig. 18-4. A pitched roof would raise the cost of this building. (Cedar Shake and Shingle Bureau)

Fig. 18-5. Water and snow will not stay on these roofs long. When snow melts, an ice dam can form on the eaves. The metal edge flashing is used to stop leaking at the edge.
(Asphalt Roofing Manufacturers Assoc.)

Flat roofs are the least costly. There is less material and labor than in other roofs. Factories and commercial buildings use flat roofs, Fig. 18-4. They get the most space for the least cost. Nearly all space under flat roofs is used.

Pitched roofs slant. Water and snow run off the roofs like that in Fig. 18-5. The space under pitched roofs is hard to use. If roof trusses are used, most attic space is wasted.

Curved roofs are strong and decorate a structure. Some interesting roof lines are shown in Fig. 18-6. Sometimes a roof and wall are one unit. Can you tell the roof from the wall in Fig. 18-7?

HOW ROOFS ARE MADE

Most roofs have three parts. They have a deck, roofing, and flashing. A rugged roof is shown in Fig. 18-8.

A

B

C

Fig. 18-6. These buildings have curved roofs. A–Arches form the roof. (Ball State University Photo Service) B–Hyperbolic curves are based on math. (The Manitowoc Co., Inc.) C–A geodesic dome. (TEMCOR)

Decks

Roof *decks* form a connecting bridge across the framing members. Decks are designed to support the roofing material.

The roof deck must be strong. It has to hold up:
- The weight of the roofing materials.
- The water and snow load.
- The roofers and their tools.
- The mechanical equipment on the roof.

Nails, clips, or glue fasten the roofing to the deck.

Roof decks are made of many materials. Concrete, boards, plywood, fiberboard, flakeboard, and metal are the most common. Two roof decks are shown in Fig. 18-9. Some of the sheet metal roofing materials do not need decking. Sometimes decking is left visible from inside if it will decorate a room with a cathedral ceiling (a peaked, open attic area).

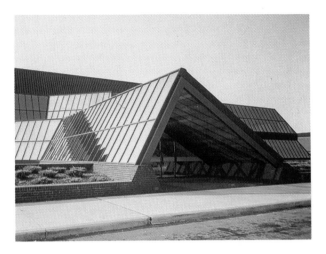

Fig. 18-7. Does this structure have walls? (Monsanto Polymer Products, Inc.)

Fig. 18-8. A roof consists of a deck, roofing, and flashing. Can you find the parts of a roof in this picture? (Owens-Corning Fiberglass Corp.)

Insulation can be added to a roof deck. Rigid panels of foam and fiber are often used. Lightweight concrete and cast gypsum are used in concrete or steel structures. These materials are foamed or have foam plastic beads in them.

Roofing

Roofing protects the building from rain and snow. Built-up, liquid, and one-ply roofing systems are used for flat roofs. Shingles and sheet roofing are used on pitched roofs.

Built-up

Built-up roofing consists of building felt, gravel, and bitumen. *Building felt* is plant fiber and asphalt. The fiber is made into a sheet. Then it is soaked with asphalt. *Bitumen* is made of asphalt or tar. Asphalt is a by-product of oil refining. Tar is the remains after coal is made into coke.

Layers of hot bitumen are covered with building felt. Each layer is called a *ply*. The more plies the

A

B

Fig. 18-9. Common deck materials. A–Flakeboard. (SENCO Products, Inc.) B–Steel decking under ''cellular glass'' slabs. (Johns-Manville Corp.)

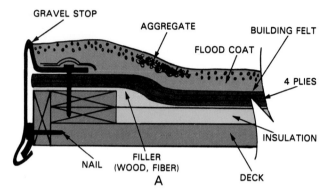

GRAVEL STOP
AGGREGATE
BUILDING FELT
FLOOD COAT
4 PLIES
INSULATION
NAIL
FILLER (WOOD, FIBER)
DECK
A

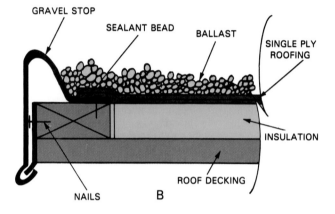

GRAVEL STOP
SEALANT BEAD
BALLAST
SINGLE PLY ROOFING
INSULATION
NAILS
ROOF DECKING
B

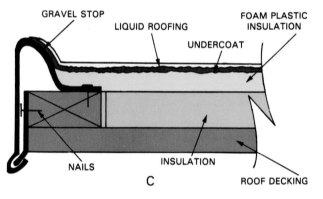

GRAVEL STOP
LIQUID ROOFING
FOAM PLASTIC INSULATION
UNDERCOAT
NAILS
INSULATION
ROOF DECKING
C

Fig. 18-10. Flat roofs are made in several ways. A–Built-up. B–One-ply member. C–Liquid.

gate (sand or gravel) is spread over it. The aggregate protects the roofing from the sun. People can walk on the roof without causing it to leak.

Liquid

Liquid roofing materials are sprayed or rolled on and allowed to harden. The first coat is a foam plastic. It smooths the surface and insulates the structure. The final coat is made of silicone rubber. Liquid roofing works well on curved roofs.

One-ply

As the name implies, one-ply roofs have only one layer. It is a sheet of special plastic put on a very smooth deck. Open places 1/8 in. wide or wider are

roof has, the longer it will last. The layers of built-up roofing are shown in Fig. 18-10.

The last layer of hot bitumen is called a *flood coat.* It is very thick. While it is still hot, an aggre-

filled. The sheets are lapped over and glued to each other and to the deck. This roofing material can be applied to any roof shape.

Shingles

Shingles are small flat or curved pieces of material fastened to the deck. They are made of asphalt, wood, slate, metal, clay, or concrete. Today, asphalt is the most common. Five kinds of shingles have been used in Fig. 18-11. Most of them are measured and sold by the square. A *square* will cover 100 sq. ft. of roof.

Asphalt shingles are laid on a solid deck (no gaps). Most decks are plywood or flakeboard. Other shingles can be put on open decking of wood. Fig. 18-12 shows open decking. Refer again to Fig. 18-9A. The roofer in Fig. 18-9A is putting on solid decking.

A

B

C

D

E

Fig. 18-11. Roof coverings include: A–Slate shingles. (Pella/Rolscreen Co.) B–Asphalt shingles. (SENCO Products, Inc.) C–Clear plastic roofing. (TEMCOR) D–Sawn red cedar shingles. (Cedar Shake and Shingle Bureau) E–Imitation clay.

Fig. 18-12. The deck for rigid roofing can be spaced.

Fig. 18-14. Staplers speed up roofing jobs. Wide staples are used so they do not tear out. (SENCO Products, Inc.)

The narrow blue strip on the left side and bottom edge of the roof in Fig. 18-13 is a ***drip edge.*** It protects the edge of the decking from water that could run under.

Shingles are attached with nails and staples. The method varies with the kind of shingle. All nails should be hot-dipped galvanized, copper, aluminum, or stainless steel. Staples can be used to fasten asphalt and wood shingles. A stapler is being used in Fig. 18-14. The worker is using nails in Fig. 18-15.

Clay and slate shingles are laid in much the same way. The details differ some for each kind of shingle.

Sheet metal

Sheet metal roofing comes in many forms. Copper, lead, and aluminum need no surface treatment. Steel must be plated or enameled. Copper is the longest lasting and the most costly. Metal is held to the deck with clips, screws, and nails, Fig. 18-16. Metal roofing is easy to install and lasts a long time.

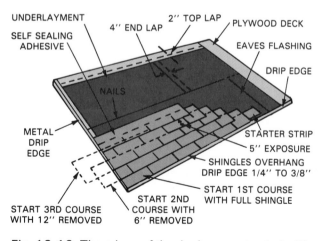
Fig. 18-13. The edges of the deck are protected with a drip edge. Shingles overlap so water cannot get in the structure.

Fig. 18-15. The roofer is putting on the cap. Hot-dipped galvanized roofing nails are used. (Asphalt Roofing Manufacturers Assoc.)

Flashing

Flashing is used to stop leaks. Leaks occur where a roof section joins a second surface. When two roof sections meet they form a valley. Flashing will prevent a leak where a roof meets a wall. Pipes and chimneys that pass through a roof are flashed. Metal, asphalt roofing, and asphalt plastic cement are used. Places where flashing is used are shown in Fig. 18-17.

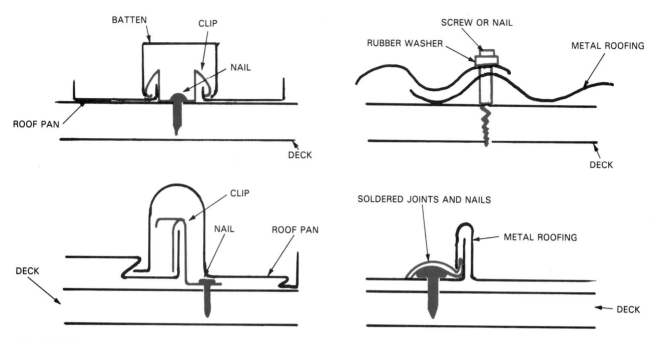

Fig. 18-16. Here are end views of four ways metal roofing is attached to a deck. Many more ways are used.

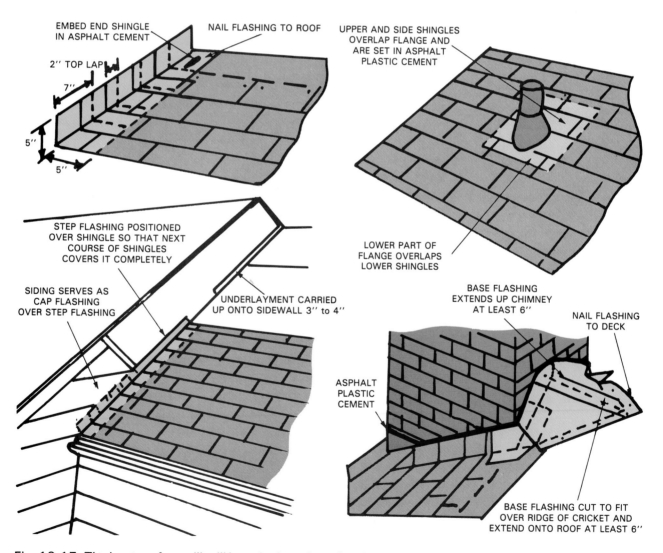

Fig. 18-17. The best roofers will still have leaks unless they install flashing. Here are four places where it is needed. (Asphalt Roofing Manufacturers Assoc.)

ENCLOSING WALLS

Walls too, are designed to keep building interiors comfortable and dry. At the same time, there must be openings that let in light and allow people to enter and leave. Walls, then, can be a combination of many different materials. The openings are closed in by windows and doors. The permanently closed sections of the wall are made of wood, masonry, and other durable materials. This chapter defines and describes curtain walls.

Installing Doors and Windows

Doors and windows are usually installed after the roof is complete. In this way, much of the traffic from roof framing work proceeds through clear door openings. Also, a completed roof covering keeps door and window frames dry.

Doors

A doorway is an opening through a building wall. The opening is fitted with a door frame. Door frames are made of wood or metal. The frame holds the hardware and presents a finished appearance. Parts of the frame are shown in Fig. 18-18. The top of the frame is the head. The bottom is the threshold. A door is attached to the frame by the hardware (hinges and latch or lock).

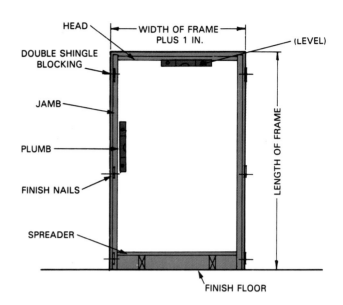

Fig. 18-18. Interior door. Door frames are set level and plumb. Nails are driven through jamb and shims into wall framing.

There are common names for door types. The doors used to enclose a structure are exterior doors. The one most used for entering is called an entrance door. Combination doors have screens and storm windows in them. Doors that resist fire are called fire doors.

Most doors swing on hinges. Doors with other operating methods are shown in Fig. 18-19.

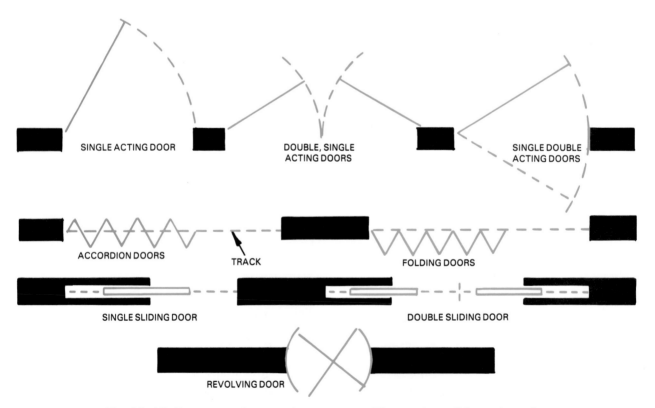

Fig. 18-19. Doors open in more than one way. They swing, slide, and revolve.

Doors can be panel or flush type. The most common type door is the panel door. Flush doors look like a slab of wood. Some flush doors are steel. Examples of both panel and flush doors are shown in Fig. 18-20.

Vision panels are used so people can see out. Double thickness glass is used today. It reduces heat loss.

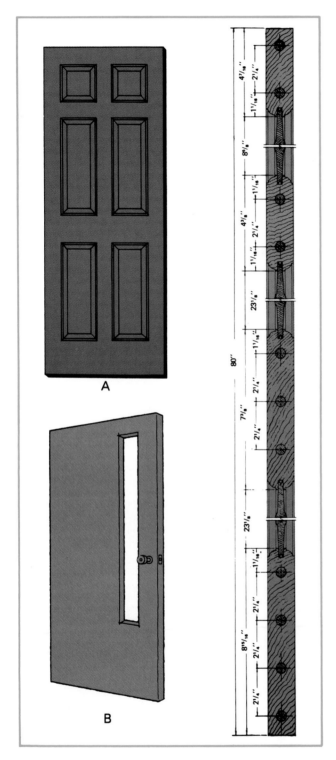

Fig. 18-20. Types of doors. A–Panel door. (Maywood, Inc.) B–Steel flush door. (Ceco Corp.)

Insulated doors are used to save heat. The core is foam plastic. Magnetic seals are used around the edges.

A door is installed in an opening in a wall frame called a rough opening. The rough opening for doors is 1 in. wider and higher than the frame. Refer back to Fig. 18-18. Shims made of sawn shingles are used to hold the door plumb. Nails are driven through the jamb and shims into the framing. Trim boards cover the gap.

Windows

A window is another opening to close up in a wall. It lets in light and fresh air.

Windows must be weathertight and be built to last. They must also be reasonably priced, easy to install, and good looking. They are made of wood and metal. The wood parts of some windows are covered with plastic and metal. Many types are made. Fig. 18-21 shows three.

The top of the window is the *head.* The *jamb* is the side. The bottom is a *sill.* A *sash* is the glass and its frame. A *mullion* is the member between two sashes.

The glass is called glazing. Double glazing and weatherstripping are used to reduce heat loss. Screens are used to keep out insects.

The rough opening for a window provides space to fit the window, Fig. 18-22A. The window is set into the opening in B. A window should be plumbed and leveled. Nails are used to hold a window in place as in C. Caulking is used to seal between the trim and siding, D. Insulation is used around the window frame inside. It is important that insulation fills the space between the window frame and wall framing.

COVERING OUTSIDE WALLS

Outside walls are enclosed or protected with many kinds of materials. Wood, stone, glass, plastic, concrete, and metal are only a few. The materials come in many forms. They include blocks, strips, slabs, plates, and panels.

Frames for buildings were designed to last a long time. The materials that cover them must last, too. Frames are enclosed in two ways: masonry and curtain walls.

Masonry Walls

Masonry walls are made of brick, concrete block, tile, or stone. Pieces of the materials are set in mortar. The framing members are encased. The

Fig. 18-21. There are many ways for windows to open. The four most common are shown. A—Double hung windows slide up and down in a track. B—Casement units swing out on hinges. C—Ventilating skylight is placed on a roof on hinges. D—Awning units open from the bottom. (Pella/Rolscreen Co.)

Fig. 18-22. The steps used to install a window are: A—Make a rough opening. B—Set in window, plumb, and level. C—Nail in place. D—Caulk outside, insulate and add trim inside. (Marvin Windows)

spaces between the members are filled and openings for doors and windows are left. Masonry walls can add to the strength of a building. The load-bearing strength of masonry walls can be important.

A wall called a *solid wall* has all spaces between the units filled with mortar. If a space between the facing and backing units is left, it is called a hollow wall. The space is for insulation, wiring, or pipes.

A wall called a *composite wall* is made up of two or more kinds of units. Face brick may be backed up with concrete block. The two walls can be tied together in many ways.

Curtain Walls

Curtain walls are a second way to enclose frames. Curtain walls are attached to the frame. They keep the forces of nature out and make the building look good. Curtain walls do not carry any of the roof or floor loads.

There are four basic kinds of curtain walls:
- Custom walls.
- Commercial walls.
- Industrial walls.
- Masonry veneer.

Custom walls are designed for one project. Fig. 18-23 shows this kind. The architect wanted the building to project a unique image.

Commercial walls are made up of standard parts. The building front in Fig. 18-24 is a commercial wall. The glass and metal channels are standard stock. They are only cut in special cases called out in the drawings.

Industrial walls are built from pieces, strips, or panels of standard material. Bricks, siding, and plywood are industrial wall stock.

A *masonry veneer wall* is used to cover the wall framing. Stone and brick are often used. Brick veneer construction is shown in Fig. 18-25.

Siding

Metal, wood, vinyl, and fiberboard strips are used to enclose buildings, Fig. 18-26. The materials come in many shapes and sizes. Fig 18-27 shows a range of wood siding shapes. Fig, 18-28 shows the use of a plastic sheet that helps keep cold air out.

Metal and vinyl strips are made to interlock. Interlocking panels have joints on their edges, Fig. 18-29.

Materials are chosen for low weight, easy installation, or for good insulation. Hardboard and aluminum strips are being installed in Fig. 18-30.

Fig. 18-23. Custom curtain walls were made special for these three structures to cover their steel frames. (Skidmore, Owings & Merrill)

Fig. 18-24. Standard stock metal and glass were used to enclose these walls. The fabricator cut them to size. The parts were assembled on the site. (Elliott Corp.)

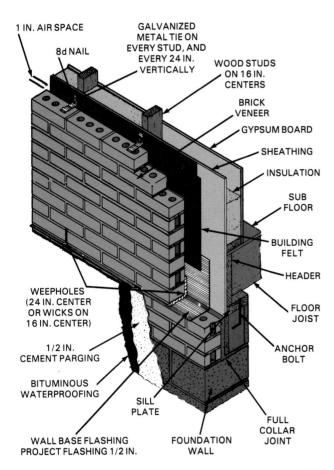

Fig. 18-25. Wood and steel frames are covered with brick or stone. Wall ties are used to hold the facing to the backing wall. (Brick Institute)

A

B

Fig. 18-26. Metal and wood are used to enclose these buildings. A—Aluminum. Quantity used is measured in squares. (Reynolds Metals Co.) B—Redwood. (California Redwood Assoc.)

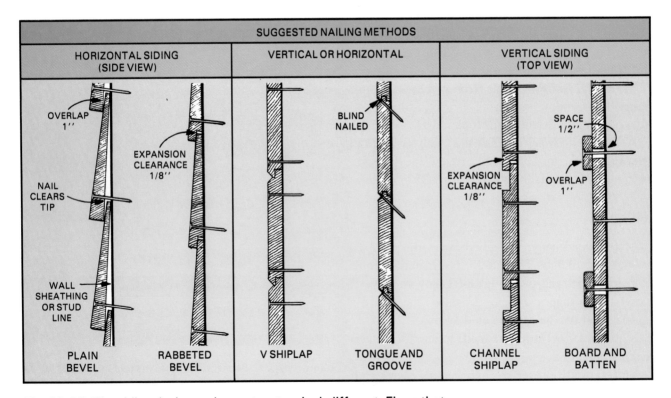

Fig. 18-27. The siding design makes a structure look different. Firms that make siding offer more than one shape. (California Redwood Assoc.)

Fig. 18-28. An air barrier is away to make a building warmer in winter. It will reduce air infiltration. That means fewer drafts affect the building.

A

B

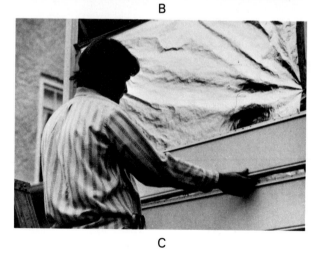

C

Fig. 18-30. Strip siding is put on from the bottom up. A–Starter strip is attached. (SENCO Products, Inc.) B–The strips are kept level. (Masonite Corp.) C–Each piece hooks onto the last. (Reynolds Metals Co.)

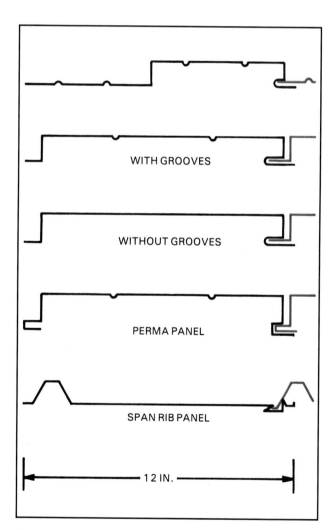

WITH GROOVES

WITHOUT GROOVES

PERMA PANEL

SPAN RIB PANEL

12 IN.

Fig. 18-29. Interlocking panels are easy to install. They are easy to fasten and they hold tight. Caulking in the joint makes the joint tighter.

A

B

C

Fig. 18-31. These curtain walls are made from panels. A–Granite. (©Cold Springs Granite Co.) B–Glass. (Monsanto Polymer Products, Inc.) C–Hardboard. (Masonite Corp.)

Aluminum is light, and hardboard helps insulate a structure.

Panels are made of stone, glass, concrete, metal, or wood. Panels may be single layers. Stone, glass, and hardboard panels were used in Fig. 18-31.

Panels can have more than one layer, Fig. 18-32. They are referred to as *composite panels.* The layers of the panel provide special features. Looks, insulation, and strength are three features. Thin metal, plywood, and rigid foam plastic may be joined. The metal resists weather and looks good. The plywood gives strength. The foam plastic keeps the building warm in the winter.

Plaster stucco is like plaster with sand in it. Expanded metal is put onto the outside wall. A scratch coat is applied first. The brown coat is put on next and the finish coat is put on last. Stucco is strong and will last a long time.

An *exterior insulating finish system (EIFS)* is an alternative to stucco and brick. It is a watertight and airtight finish that looks like stucco. EIFS requires a watertight wrap over the sheathing. The joints are overlapped or taped with a water barrier tape. Starter strips and beading strips are used at the edges and joints. Stainless steel mesh furring strips are placed on each vertical framing member. The entire outside surface is covered with 1/2″ thick cement board. Finally, two layers of synthetic stucco are applied with a trowel.

Fig. 18-32. These composite panels have exterior plywood siding on the outside. A rigid foam insulation is the core. The interior surface is 1/2″ flakeboard, which is a strong flat base for gypsum board. (J-DECK Building Systems)

SUMMARY

Structures are enclosed to keep weather out. Roofs enclose the tops of buildings. They consist of decks, roofing, and flashing. Doors and windows provide openings in walls. Outside walls are covered with masonry and curtain walls. Masonry walls can be load-bearing. Most curtain walls are not load-bearing.

KEY WORDS

All of the following words have been used in this chapter. Do you know their meaning?

Bitumen
Building felt
Commercial wall
Composite panels
Curtain wall
Custom wall
Deck
Drip edge
Exterior insulating finish system (EIFS)
Flashing
Flood coat
Head
Industrial wall
Jamb
Masonry veneer wall
Masonry wall
Mullion
Plaster stucco
Ply
Sash
Shingle
Sill
Square

TEST YOUR KNOWLEDGE

Write your answers on a separate sheet of paper. Do not write in this book.

1. A _____-_____ roof consists of several plies of building felt with gravel and bitumen on top.
2. A "square" of shingles will cover _____ sq. ft. of roof.

Matching question: On a separate sheet of paper, match the definition in the left-hand column with the correct term in the right-hand column.

3. _____ Formed when two roof sections meet.	a.	Mullion.
4. _____ The side of a door or window frame.	b.	Flashing.
5. _____ Will surround the finish frame of a door or window.	c. d.	Valley. Head.
6. _____ Metal used to stop roof leaks.	e. f.	Jamb. Rough opening.
7. _____ A member between two sashes.		
8. _____ The top of a door frame.		

9. Curtain walls support roof or floor loads. True or false?
10. The second coat of stucco work is the _____ coat.

ACTIVITIES

1. Find out what kind of roofing your home has. If possible, look at it. How was it installed?
2. Visit a building supply store. Find out what kind of roofing materials they handle. Get brochures on how to install them.
3. Visit a roofer. Ask what kinds of roofs they install and how they do it. If you can, visit an actual work site. Record what you see. Report to the class.
4. Do numbers 1-3 with windows and doors.
5. Do numbers 1-3 with siding.
6. While downtown, see if you can find a variety of masonry and curtain walls.

Section 6
INSTALLING SYSTEMS

Chapter 19

INSTALLING CONVEYANCE SYSTEMS

After studying this chapter, you will be able to:
☐ Discuss why a stairway is put in before plumbing and climate control systems.
☐ Recall rules for design of safe stairways.
☐ Explain how two types of elevators work.
☐ Describe moving stairways, pneumatic tubes, dumbwaiters, and chutes.

The foundation and frames provide the strength of the structure. Enclosing the outside keeps out the forces of nature. Enclosing the inside makes the structure look nice and also makes it easy to clean.

The mechanical systems provide warmth, light, information, and fresh air. With these systems, people are able to live and work in a structure. Mechanical systems include conveyance (traffic movement), climate control, plumbing, electrical power, and communication systems.

Much of the mechanical system is placed in floors, walls, and ceilings. This part of the work is called *roughing-in.*

The bulky and rigid parts are roughed in first. The most flexible are roughed in last. Conveyance systems are the first to be installed. They are used to move people and things throughout the structure. Ductwork for climate control systems follows. Large plumbing pipes are roughed in at this time, too. The smaller water supply and drain lines are next. Finally, the electrical wire and communication cables are installed.

CONVEYANCE SYSTEMS

Conveyance systems make structures useful. People and supplies are moved throughout buildings with conveyance equipment. See Fig. 19-1.

All structures over one story high require upward and downward conveyance. See Fig. 19-2. Stair-ways, elevators, moving stairs, and moving walkways carry people. Freight is moved with elevators and conveyors. Papers and small items may be moved through pneumatic tubes and chutes.

Stairways

The stairway is the basic way to move people in or onto structures. See Fig. 19-3. City codes describe the width of stairways for public buildings. They may vary widely. Some common guides follow:
- The enclosure for two or more stories must be fireproof. Private homes are not included.
- Buildings with more than three floors need two stairway exits.
- The size of stairways and the distance to them is determined by the number of people on the floor.

In a home the following guides are useful:
- A main stairway should allow at least two people to pass.
- A stairway should be wide enough to move furniture up and down. A width of 3 ft. is suggested.
- Stairways should have railings.

A complete set of drawings includes stairway details. Fig. 19-4 is a drawing that shows wood stairway parts. All stairway openings appear on floor plans. New construction may require temporary stairways. An example of one is shown in Fig. 19-5.

Stairways are made in many shapes. Drawings of six shapes are given in Fig. 19-6.

Elevators

Elevators are machines that raise and lower people and freight in a structure. Elevators were invented when cities became crowded. Structures had to be built taller. Elevators were needed to build and occupy them. See Fig. 19-7. They bring material

Fig. 19-1. People and freight are moved with elevators. (AMCO Elevators, Inc.)

and workers to the workplace. After the structure is complete, elevators move people to the floors where they live and work. Today's elevators are both fast and safe.

Designing

Architect/engineers decide on the number, size, speed, and kind of elevators. Codes only state safety requirements. The A/Es have to design the system within those limitations.

Care is used in designing elevators. The speed and convenience in moving people can make or break a building's budget. The number, speed, and size of elevators depends on how they are to be used.

When elevator equipment, along with programmed controlling, is provided, the package is called *elevator service.* Elevator service is planned to be convenient and to save floor space. People need to get to the upper floors quickly. Adding more elevators does not always solve the problem.

Fig. 19-2. These moving stairways carry an even flow of people. People do not have to wait. (BIRDAIR,Inc,)

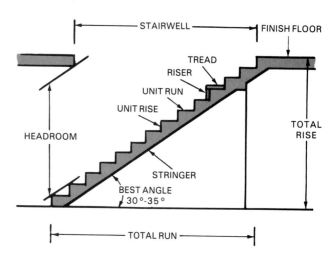

Fig. 19-4. Each part of a stairway has a name.

A

B

C

D

Fig. 19-3. Stairways are used in these projects. A—Spiral stairs mount onto tank. (Combustion Engineering) B—Stairway has safety enclosure. (Texaco Inc.) C—Poles help support stairway and platform. (Universal Tank and Iron Works, Inc.) D—Part of stairway uses existing machinery housing as base. (Stran Buildings)

Fig. 19-5. This temporary stairway was installed early in construction for use by the workers. When construction is complete, damaged parts will be replaced before the floor covering is installed.

Each hoist shaft takes up floor space. This space cannot be rented when it is used for elevators.

A/Es have worked out many ways to provide elevator service. Four service plans are shown in Fig. 19-8. The first plan, A, is for low buildings. All elevators stop at each floor. In the second service plan, B, elevators stop only at certain floors. The third plan, C, is for very tall buildings. Sky lobbies are used. The *sky lobby* is a transfer point where people change elevators. Shops are set up in the sky

Fig. 19-7. Elevators move people and freight. Without them tall buildings would be nearly worthless. (Dover Elevators)

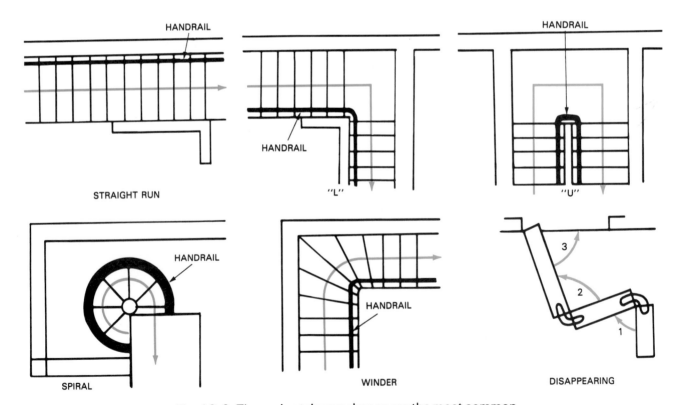

Fig. 19-6. These six stairway shapes are the most common.

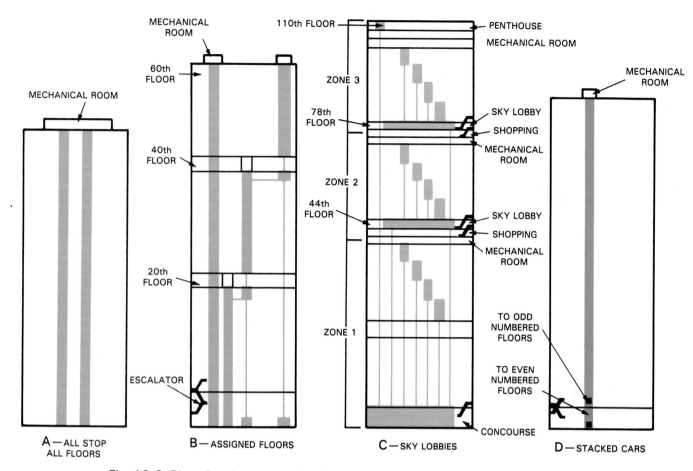

MECHANICAL ROOM

MECHANICAL ROOM

110th FLOOR — PENTHOUSE
MECHANICAL ROOM

60th FLOOR

ZONE 3

MECHANICAL ROOM

78th FLOOR — SKY LOBBY
— SHOPPING
— MECHANICAL ROOM

40th FLOOR

ZONE 2

44th FLOOR — SKY LOBBY
— SHOPPING
— MECHANICAL ROOM

20th FLOOR

ZONE 1

TO ODD NUMBERED FLOORS

TO EVEN NUMBERED FLOORS

ESCALATOR

CONCOURSE

A—ALL STOP ALL FLOORS   B—ASSIGNED FLOORS   C—SKY LOBBIES   D—STACKED CARS

Fig. 19-8. Plans for elevator service have been worked out to move people quickly.

lobbies. The fourth plan, D, uses double cars. The bottom car stops at odd numbered floors. The top is for the even numbered floors.

Uses and special kinds

Elevators in buildings move people and supplies. *Passenger elevators* get people to the floors with their homes or workplaces. Freight is hauled on *service elevators.* They are often larger than passenger elevators. They carry heavy loads and move slower. Some buildings have extra large freight elevators. They can raise and lower loaded truck trailers.

Elevators are moved with mechanical and hydraulic power. See Fig. 19-9. *Mechanical elevators* use cables to move the car. They are used in high-rise buildings. *Hydraulic elevators* use high pressure oil. They are quiet and smooth running. They are used only in low-rise buildings.

Operation

Elevators have many parts. Refer back to Fig. 19-9 for a view of these parts. The part most people see is the *car.* The car hangs from *cables* which go over a *sheave.* Through a pulley system, one cable holds a second cable for a *counterbalanced weight.*

The weight helps the motor lift the car and people. A second weight is on the car. A third cable is attached to the bottom of the car. It passes through the bottom sheave and hooks to the weight.

Many years ago, all elevators were run by a person in the car. Modern elevators need no operators. They are run by a *control system.* See Fig. 19-10. It keeps track of all calls for the elevators. The control system "decides" who should be picked up first. The people are then taken to their floors.

A dispatcher in the lobby often watches the operation of the elevator. This person is on duty if anything goes wrong. Under normal conditions, the computer control system runs the elevators better than people can.

Installing

Elevators and movable stairways are assembled and installed by persons or firms called *elevator constructors.* They also service and repair them. The contractor does the "roughing-in." They install the hoist way, floors, and beams that are needed. Guide rails, hoisting machinery, hydraulic systems, cars, and controls are installed by the elevator constructor. Careful testing is required. See Fig. 19-11.

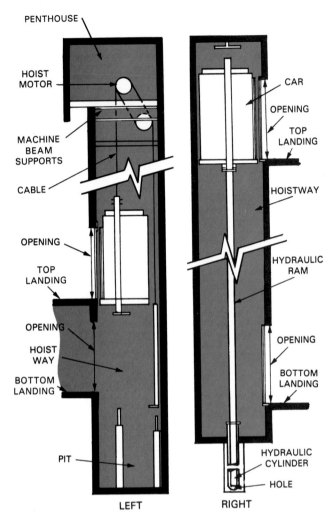

Fig. 19-9. Both mechanical and hydraulic power are used for elevators. Left. Mechanical elevators have a penthouse. Right. A deep hole is dug for hydraulic elevators. (Westinghouse Elevator Co.)

Fig. 19-10. Elevator control systems can decide when to stop at a floor. Control panels use electronic circuits. (U.S. Elevator)

Fig. 19-11. Controls for elevators are installed by elevator installers. (U.S. Elevators)

Most elevator constructors work for elevator manufacturers. Moving stairways are also built by those who work for the manufacturer.

Moving Stairs

Moving stairs can move many people and keep them moving. People do not have to wait.

Moving stairs vary in width and in the speed they travel. They are set up in patterns. Fig. 19-12 shows two common patterns.

The moving stairway has steps, a chain, wheels, a track, handrails, and a drive unit. *Steps* are attached to a wide chain. The *chain* has wheels that run in a track. *Handrails* move with the steps to make it safer to ride. The *drive unit* pulls the chain along.

Material Handling Equipment

When constructing a building for a factory, an important part of the structure is a conveyance system. This can include equipment for moving supplies or finished products within the factory. An assembly line is the most common arrangement for material handling. The line may or may not be manually monitored at multiple workstations. Fig. 19-13 shows material handling systems that operate automatically.

Parts for building many conveying systems are standardized. The parts can be purchased from an equipment supplier.

Fig. 19-12. Moving stairways are installed in several patterns. Separate, crisscross, or parallel patterns are most common. (Westinghouse Elevator Co.)

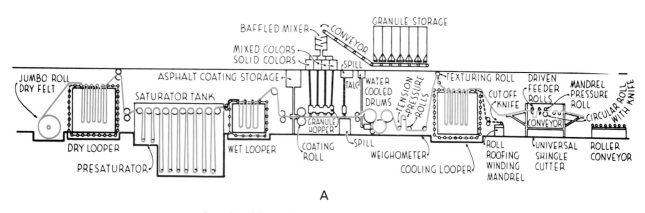

A

Lumber Manufacture at a Typical Sawmill

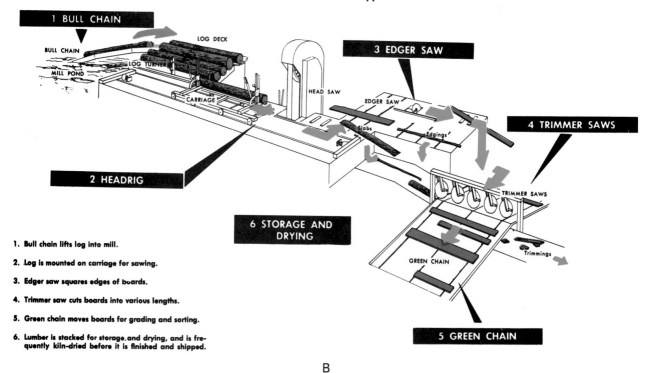

1. Bull chain lifts log into mill.

2. Log is mounted on carriage for sawing.

3. Edger saw squares edges of boards.

4. Trimmer saw cuts boards into various lengths.

5. Green chain moves boards for grading and sorting.

6. Lumber is stacked for storage and drying, and is frequently kiln-dried before it is finished and shipped.

B

Fig. 19-13. Material handling systems. A–System for making rolls of asphalt roofing felt. (Asphalt Roofing Manufacturers Assoc.) B–How lumber is produced at a saw mill. Very little help is needed from workers. (Forest Products Laboratory)

Conveyors

Conveyors carry paper, mail, and small items throughout a structure. See Fig. 19-14. They run horizontally and vertically. The U.S. Post Office has urged architects to include them. Conveyors reduce the work in delivering mail.

Most conveyors have chains running on sprocket wheels. Trays are fastened to the chains. Each tray has a sensing and controlling device on it. The device is set to push the tray off at the desired floor.

Horizontal conveyors are used in some airports. They move people and luggage long distances. People step onto a moving belt and set their luggage beside them. A moving handrail helps people keep their balance.

Dumbwaiter

A *dumbwaiter* is a small freight elevator. Fig. 19-15 is a picture of one. It is run manually. A dumbwaiter can carry more weight than a conveyor, but it is not automatic.

Pneumatic Tubes

Pneumatic tubes are used to carry small lightweight items. See Fig. 19-16. The item is placed in a bullet shaped case. The case is put into the inlet of a

Fig. 19-15. Dumbwaiters move freight in buildings. This one loads itself. Most can be manually operated from a panel. (Courion Industries, Inc.)

tube. Moving air in the tube pushes the case to its destination quickly.

Chutes

A *chute* is a round or square tube that runs between floors. When items are placed in the chute, gravity pulls them down. Chutes are used in apartment buildings to dispose of garbage. Some homes

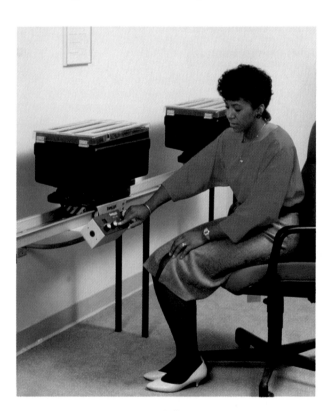

Fig. 19-14. Conveyors save delivery time in this mini-mail-room. (TransLogic Corp.)

Fig. 19-16. Medicine is delivered quickly at this health care center. Air blows the cartridge through the tube. (Translogic Corp.)

have clothes chutes for dropping dirty clothes from the upstairs to the basement.

SUMMARY

Efficient structures have conveyance systems that are easy to use. They move people and supplies. Stairways, elevators, conveyors, dumbwaiters, pneumatic tubes, and chutes are used. They are selected by architects and designed by engineers. The capacity, speed, and kind needed guide the engineer in designing the system. Elevator cars and hoist equipment are installed by elevator constructors.

KEY WORDS

All the following words have been used in this chapter. Do you know their meaning?

Cable
Car
Chain
Chute
Control system
Conveyance system
Conveyor
Counterbalanced weight
Drive unit
Dumbwaiter
Elevator
Elevator constructor
Elevator service
Handrail
Horizontal conveyor
Hydraulic elevator
Mechanical elevator
Moving stairs
Passenger elevator
Pneumatic tube
Roughing-in
Service elevator
Sheave
Sky lobby
Steps

TEST YOUR KNOWLEDGE

Write your answers on a separate sheet of paper. Do not write in this book.
1. The most basic method to move people and things in a structure is a _____.
2. List seven methods used to move people and things throughout a structure.
3. When is a conveyance system roughed-in?
4. Machines that move people or freight within a structure are called _____ and _____.
5. The two kinds of elevators described are called _____ and _____.
6. A smooth flow of people can be moved from floor to floor with a/an _____.
7. A system that uses moving trays and chains to move small items through a building is called a/an _____.
8. A small manually operated freight elevator is a/an _____.

Matching questions: On a separate sheet of paper, match the definition in the left-hand column with the correct term in the right-hand column.

9. _____ Involves conveyance and climate control.
10. _____ Describes the width and number of stairways to use in public buildings.
11. _____ Names of stairway types.
12. _____ Can reach both the even and odd numbered floors.
13. _____ Helps a motor lift a car and the people.
14. _____ Common factory conveying arrangement.
15. _____ Holds small items for a chain drive conveyor.
16. _____ Message moves inside of this.

a. Counter balance.
b. Winder, "U," "L," straight run.
c. Pneumatic tube.
d. Tray.
e. Assembly line.
f. Mechanical system.
g. City code.
h. Stacked cars.

ACTIVITIES

1. Make a model of an elevator shaft out of balsa wood. Build the car and a pulley system. Make and adjust weights for the car and for a counterbalance.
2. What methods of conveyance are used in your school? Your home? List and describe them.
3. At a library, find out further details about moving stairways. How wide is the drive chain, and what metal is used for the chain, wheels, and steps? How much horsepower do most drive motors need?

Adjusting an elevator door motor in an elevator shaft. (Dover Elevators)

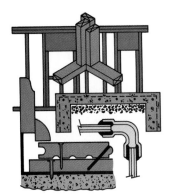

Chapter 20

ENERGY AND CLIMATE CONTROL

After studying this chapter, you will be able to:
☐ Discuss five sources of energy.
☐ Describe how ductwork is installed.
☐ Define active and passive solar energy.
☐ List uses for heat pumps, air conditioners, and humidifiers.
☐ Explain the operation of thermostats and zone controls.

Climate means the temperature, humidity, purity, and circulation of the air inside structures. Four systems are used to maintain air quality. Heating and cooling systems control the temperature. Humidifiers and evaporative coolers add moisture to air. Moisture is removed with dehumidifiers and ventilators. Air exchangers replace stale inside air with fresh outside air. Filters and electronic air cleaners remove dust from the air.

CHANGING TEMPERATURE

Heat can be added or removed from air in enclosed structures. Many methods have been devised to maintain a comfortable temperature. All of them consist of the following parts: an energy source, a converter, a distribution system, and controls.

Energy Sources

In the past, wood and coal were the primary sources used to produce heat. Wood is still being used where it is easy to get. Coal is plentiful. Its main use is to generate electricity. It is seldom used anymore to heat space.

Fuel oil is made from crude oil. It is easy to transport and provides a lot of heat for its weight. No ashes remain after it is burned. Fuel oil is easier to burn and control than wood or coal. Wood or coal require special handling and a large storage space close to where they are used.

Gas is also very easy to use. It burns cleanly, requires no handling, and provides an even heat. Gas can be compressed to a liquid and hauled in a tank. It is often stored where it is used, Fig. 20-1. But, most often, it comes through buried pipes as a

Fig. 20-1. One important construction job is installing tanks for liquefied gas. When oil is used, it is often stored in underground tanks.

gas. The gas from the pipe is supplied at a standard pressure. A *regulator* lowers the pressure before it is used. See Fig. 20-2. This device is placed outside the building it serves. A gas meter keeps track of the amount of gas used.

Another form of energy is electricity. Electricity does not require storage or pipes. It is carried by several wires. Electricity is moved over long distances at a high voltage. Near the point of use, the voltage is lowered with a *transformer.* A meter shows how much electricity is used.

The fuels described above are being used up. Gas, oil, and coal are exhaustible. That means that once they are used they cannot be replaced. It took millions of years to make these fuels. We cannot wait that long to get more. Wood is renewable, but we need too much. There is not enough land to grow the trees we would need for energy.

More and more, the sun is being used for heat. The sun's light is called *solar energy.* It is inexhaustible. The sun gives off energy that is free. There is enough to serve our needs. The problem is to collect, convert, and distribute it. Better equipment needs to be designed and built.

Air and water have heat in them. Even when they are cold they hold some heat. This heat can be removed and used.

Converters

Energy is used to heat and cool the air in buildings. Converters are used to change other energy forms for use in heating or cooling. Converters may use electricity, fuels, and solar energy as sources of energy.

Heating

Heat converters are used to warm air and water. They consist of a heat producer and a heat exchanger. Wood, coal, gas, and oil are burned. The heat producer for these fuels is called a *combustion chamber.* In the combustion chamber, fuel and air are burned. See Fig. 20-3. Combustion chambers are designed to contain the burning and exhaust the fumes.

A

B

Fig. 20-2. A—Homes have small gas regulators to reduce the pressures both from tanks and the gas company. B—Much larger regulators are used in factories.

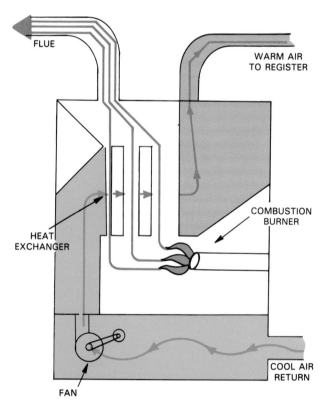

Fig. 20-3. Heat from combustion is transferred to the air that flows over the combustion chamber.

After heat is produced, it is distributed within the structure by ducts. A blower fan is used. The heat also distributes itself by convection (natural rising and falling). See Fig. 20-4.

Coal and wood need a combustion chamber. Air comes in, the fuel burns, and fumes go out. Oil and gas need a burner. A *burner* is designed to mix the fuel and air.

A *heat exchanger* takes heat from the hot gases in the combustion chamber. See Fig. 20-5. Water or air is moved over the heat exchanger. The heat moves into the water or air.

Efficient converters have efficient heat exchangers. They remove most of the heat from the hot gases. New furnaces, Fig. 20-6, remove up to 97 percent of the heat of combustion. This is nearly twice as much heat as most older furnaces remove.

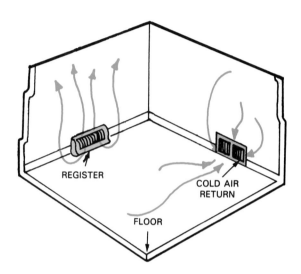

Fig. 20-4. Some steps in heating a building depend on the principle of convection. Warm air from a floor-level register rises, heating the room. Cool air enters a low opening that takes it back to the furnace.

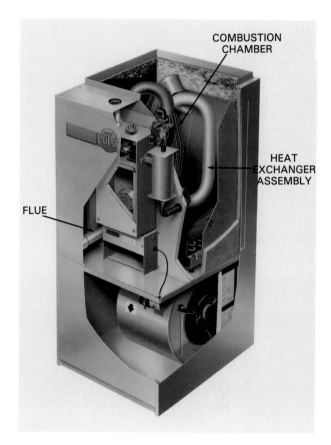

Fig. 20-6. A pulse furnace does not require a pilot light, burner, or chimney. The heat exchanger is very efficient. (Lennox Industries, Inc.)

Fig. 20-5. In this high efficiency furnace module, a special fluid is used to collect the heat. (Amana Refrigeration, Inc.)

Electric heaters use *heating elements*. Elements have resistance wires in them which are the heat producers. Electricity flows through the wires. The wires are designed to resist the flow of the electricity. When electricity flows through a resistance, it gives off heat. The more electricity flowing through the element the hotter the wire gets.

Electric elements are used to heat liquids and air. A heat exchanger is needed.

There are no hot gases with electric heat. There is an electric current, though. The electric current has to be insulated from the liquid or air, or the user could receive a shock. Insulation also prevents an electrical short circuit that could burn out the heating element.

Solar collectors

Solar collectors are used to convert sunlight into heat. See Fig. 20-7. A black surface is turned towards the sun. This is the convertor. Black absorbs more of the energy from the sun than other colors. The black surface is covered with glass. Most of the sunlight passes through the glass and hits the black surface. It is changed from light waves to heat waves. Heat waves do not pass through glass as easily as light waves. Therefore, the heat gets trapped inside the collector.

There are active and passive collectors. *Active collectors* have a liquid or air moving through them. This is the heat exchanger. The water or air absorbs the heat. It is used to heat something or is carried to a heat storage unit.

Passive collectors are usually combined with large masses of material. See Fig. 20-8. The mass of material can store heat from the sun. It may be a thick concrete wall or barrels of water. Heat is given off to the air when the air around the mass is cooler than the mass.

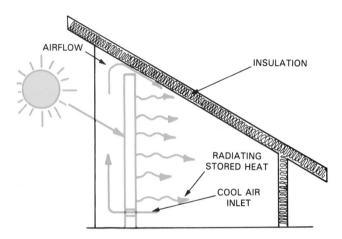

Fig. 20-8. Passive solar construction. Thick walls of concrete or tanks of water absorb heat. They give it up when the surrounding air cools.

Cooling

Air conditioners remove heat from the air in a structure. The heat is released outside the building. An air conditioner also removes water vapor. The water vapor is turned into liquid water and drained outside.

Gas units can be used for cooling, but electricity is more popular. *Central air conditioning* units are used to cool an entire building. Window air conditioners cool only a room or two.

Two separate currents of air move through an air conditioner. See Fig. 20-9. There is hot, humid air from the room. It passes across cold evaporator

A

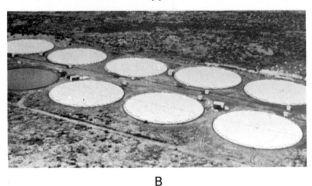

B

Fig. 20-7. These solar collectors convert sunlight into heat. A–The heat is used to warm water. (Reynolds Metals Co.) B–Large system produces both heat and electricity. (United Energy Corp.)

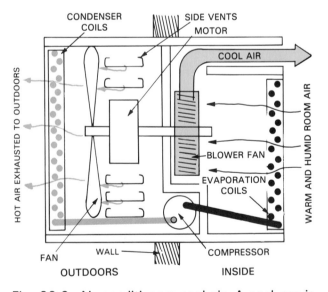

Fig. 20-9. Air conditioners cool air. A cool gas is warmed with inside air. A compressor puts it under pressure until it becomes a hot fluid. Outside air is blown over pipes to cool the hot fluid. A valve lets it cool by evaporating again. Inside air is blown over the pipes that are cooled. The cycle continues.

coils. Here it is cooled and dehumidified (water taken out). It is then blown back into the room.

At the same time, air from the outside is blown over hot condenser coils. These coils contain **refrigerant**. The refrigerant was heated by the inside air. The outside air cools the refrigerant. A compressor is the heat producer. The evaporator coils and the condenser coils are the heat exchangers.

Evaporative coolers, Fig. 20-10 cool and moisten air. They have a mat of loose material. A trickle of water keeps the material wet. A fan pulls outside air through the wet mat. The air is cooled and moisture is added. The cool, moist air is added to the enclosed space. Evaporative coolers are used mostly by people who live in a hot, dry climate. The hotter and drier the outside air, the better evaporative coolers work.

Heat pumps can heat or cool a home. See Fig. 20-11. When it is hot inside, the heat pump works as an air conditioner. It removes heat from the inside. The machine works in reverse in cold weather. Heat from the outside air is removed and used to heat the inside. Heat pumps work until the outside air drops to around 20°F (7°C).

Fig. 20-11. Heat pumps take heat from one place and put it in another. It can put heat into a building or remove it. All the owner has to do is to change the setting. (Lennox Industries, Inc.)

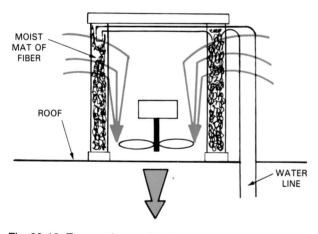

Fig. 20-10. Evaporative coolers both cool and humidify the air. Air is drawn through a moist mat.

DISTRIBUTION SYSTEMS

Distribution systems move heated or cooled water or air throughout a building. Heat can be moved in four ways. *Pipes* are used to move heated water. Air is forced through *ducts*. Some electric heating systems use *wires*. Stoves and passive solar heating systems use *gravity* and *radiators*.

Pipe Systems

Converters that warm water or make steam have boilers. The **boiler** is used to store the hot water or steam. A pump is used to force the water through the pipes.

In the room is a convector. A **convector** is warmed by the water or steam. Air begins to move past the convector. As the air moves by, it picks up heat. Some convectors have fans to help move the air.

After passing through the convector, the water is piped back to the boiler and is heated again. The hot water and steam mixture is stored in the boiler until it is needed. Chilled water can be pumped through convectors, too. It is used to cool a room.

The pipes and boilers are covered with insulation. See Fig. 20-12. **Insulation** helps maintain the correct temperature of the water or steam.

Another heating method uses a radiator. Radiators are placed along the wall. A **radiator** gives direct radiation like the sun. Some ways in which they act like the sun are:

• You feel warm when you stand next to them.
• When someone gets between you and the radiator you do not feel the heat. It is like standing in the shade on a sunny day.

Pipes embedded in plaster also radiate heat. Radiant heat panels are placed in floors, walls, or ceilings.

Duct Systems

Duct systems consist of a plenum, registers, and ducts. A *plenum* is a chamber that joins several ducts to an air inlet or outlet. See Fig. 20-13. The conditioned air enters the room through registers

Fig. 20-12. Heat loss is reduced with insulation. Tape will hold the covering tight and in place. (CertainTeed Corp.)

Fig. 20-14. A fiberglass duct system is being installed. It reduces noise and heat loss. (CertainTeed Corp.)

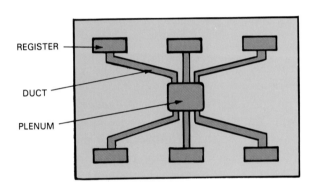

Fig. 20-13. The radial layout costs the least to install. Dampers are used to balance the system. Nearby lines are partly closed down. Long lines are left open.

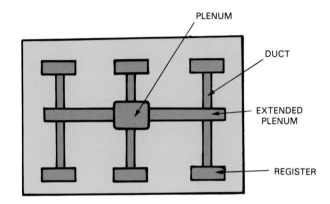

Fig. 20-15. The extended plenum looks and works better than a radial pattern. Branch ducts can be placed between floor joists.

placed in each room. The *register* spreads the air into the room.

Ducts are made of thin sheet metal, fiberglass, or plastic. They are round or rectangular in shape. Insulation keeps air cool or warm and lowers the noise of moving air. See Fig. 20-14.

There are many ways to lay out ductwork. The two common ones are radial and the extended plenum. In the *radial system,* registers are placed on the outside of rooms. Look again at Fig. 20-13. A duct from the furnace is run to each register. The shortest route possible is used. The return air is fed back through registers on inside walls.

The *extended plenum system* has a trunk line. See Fig. 20-15. It may be rectangular or round. Small ducts branch off to each register.

The plenum is usually a metal box attached to the furnace. Holes are cut in the plenum where ductwork connects.

Wires

Electric radiant heat is similar to radiant hot water. Resistance wires instead of pipes are embedded in walls, floors, or ceilings. *Resistance wires* get hot when electricity passes through them. They warm the surface, which radiates heat to people and objects in the room.

This system requires no ducts or chimneys. There are no moving parts to wear out. It is easy to install. However, it is somewhat costly to run. The structure should be well insulated to conserve the heat.

Gravity Systems

Heated air rises, cold air sinks. This is the rule that controls *gravity systems.* See Fig. 20-16. When wood is burned in a stove, the stove gets warm and heats the air around it. Warm air rises and cool air

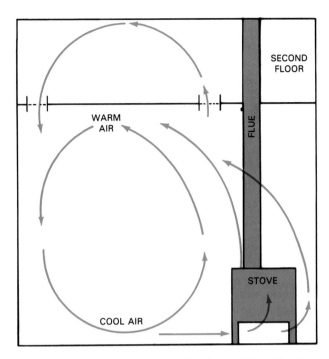

Fig. 20-16. Warm air rises. Cool air sinks. The hot stove keeps air moving.

Fig. 20-17. This is an air conditioner. The panel on the right controls the speed of the fan. The thermostat that controls the temperature is in the lower right corner. (Amana Refrigeration, Inc.)

replaces it. Openings in the ceiling will let warm air enter the upstairs rooms. Openings in the floor where air is the coolest will let the cold air drop. Gentle air currents start to flow. With careful placement of openings, two or more rooms can be heated.

CONTROLLING TEMPERATURE

Temperature controls are called *thermostats.* The main control is the room thermostat. A different thermostat controls the burning temperature in the furnace. A room thermostat turns the furnace on when the room cools. The furnace is turned off when the room gets warm.

Cooling thermostats control the air conditioner. See Fig. 20-17. The air conditioner is turned on when a room gets too warm. After the room is cooled, the thermostat turns off the power for the air conditioner.

A *zone control system* controls parts of a building. A separate thermostat controls each section or room.

A timer thermostat will automatically turn the heat up or down. In the winter, the heat is turned down when space is not used. It is turned up when it is in use. In the summer, timer thermostats work in reverse. Once the thermostat is set, it is a carefree system.

New buildings have a system of temperature sensors. A *temperature sensor* can detect small changes in temperature. A change may occur when sun shines on the side of a building. A computer adjusts the temperature automatically.

CHANGING HUMIDITY

Humidity is more correctly called relative humidity. *Relative humidity* is the amount of water vapor in the air. When air is full of water vapor, it has 100 percent relative humidity. The air feels very damp. If there is no water in the air, there is 0 percent relative humidity. A relative humidity of 30-50 percent is preferred.

In the northern United States we may want to lower humidity in summer. This is called *dehumidifying.* In winter we may want to raise it. We raise it by *humidifying* the air.

Dehumidifying Air

You can dehumidify a building by removing moisture from air. A dehumidifier can be used. A second way is to ventilate the space.

Dehumidifiers

Dehumidifiers draw moist air over cool coils. See Fig. 20-18. Water condenses on the coils and runs off into a tray or to a drain. The drier air is returned to the room. In time, the relative humidity of the air is lowered.

The addition of a *humidistat* will control the system. When the humidity gets low enough, the humidistat turns the dehumidifier off. Humidistats are set by the owner. The dehumidifier will now operate automatically.

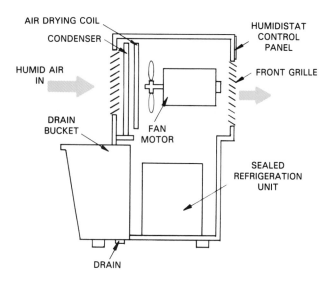

Fig. 20-18. Dehumidifiers remove moisture from the air. Cool coils condense the water in the air.

Ventilating

Air can become too moist, too warm, or polluted inside a structure. Fans are used to replace the moist, hot, or polluted air with drier, cooler, or cleaner air.

Fans can be controlled manually. The unit is turned off and on when it is needed. The humidistat or thermostat may be used to make the fan run automatically. A humidistat is used when the fan is used to reduce humidity. A thermostat is used when the space is being cooled.

When you exhaust air in the winter, you lose heat. You lose cool air in the summer if the space is air conditioned. This is wasteful. *Air exchangers* are used to save over 75 percent of the heat. Fig. 20-19 shows how they work.

Humidifying air

When air gets too dry we are uncomfortable. Sometimes our skin dries out and cracks. When the weather is cold, it is harder to stay warm in dry air than in moist air. Moisture is added to air with a *humidifier.*

Room humidifiers add moisture to each room. Central humidifiers are a part of the furnace. With central humidifiers, moist air is added to the entire structure and ductwork distributes the air.

CLEANING AIR

Dirt, pollen, and smoke pollute air. Special machines are used to purify the air.

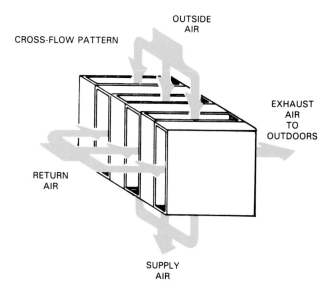

Fig. 20-19. An air exchanger saves heat when bringing in cold fresh air. The outside air runs in a cross-flow direction to the exhaust air. The outside air never mixes with the exhaust air. The cool air is heated by the warm air.
(Des Champs Laboratories, Inc.)

Filters

Filters remove a lot of the dust. Dirty air is pulled through a filter in the air distribution system. Some of the particles of dirt are trapped in the filter, which is replaced or cleaned when it is full.

Electronic Air Cleaners

Electronic air cleaners are used to remove pollen and smoke. Self-contained units or central systems can be bought. Air is blown through the air cleaners. Electric charges throw the dirt against collection plates. The dirt sticks on the plates until it is washed off. These air cleaners remove up to 90 percent of the dirt in the air.

Gases and odors are removed with fans or air exchangers. These machines were described earlier in this chapter.

SUMMARY

Climate control systems provide pure air at the best temperature and relative humidity. The temperature is adjusted with heating and cooling systems. A thermostat controls the temperature. Humidifiers, dehumidifiers, and ventilators adjust the moisture in the air. A humidistat controls the relative humidity. Filters and electronic air cleaners purify the air.

KEY WORDS

All of the following words have been used in this chapter. Do you know their meaning?

Active collector
Air conditioner
Air exchanger
Boiler
Burner
Central air conditioning
Combustion chamber
Compressor
Convector
Evaporative cooler
Dehumidifier
Dehumidifying
Duct
Duct system
Electronic air cleaner
Extended plenum system
Filter
Gravity system
Heat exchanger
Heat pump
Heating element
Humidifier
Humidifying
Humidistat
Insulation
Passive collector
Plenum
Radial system
Radiator
Refrigerant
Register
Regulator
Relative humidity
Resistance wire
Solar energy
Temperature sensor
Thermostats
Transformer
Zone control system

TEST YOUR KNOWLEDGE

Write your answers on a separate sheet of paper. Do not write in this book.

1. List three ways you can change the climate inside a structure.

2. List and give an example of the four parts of a temperature control system.

Matching questions: On a separate sheet of paper, match the definition in the left-hand column with the correct term in the right-hand column.

3. _____ Takes outdoor heat to release it inside.
4. _____ Device that makes steam.
5. _____ Joins ducts to air inlet or outlet.
6. _____ Units produce heat electrically.
7. _____ Unit saves heat when bringing in outside air.

    a. Plenum.
    b. Resistance wires.
    c. Air exchanger.
    d. Boiler.
    e. Heat pump.

8. List three sources of energy. Give one advantage and one disadvantage of each.
9. A device used to change energy for heating and cooling is called an _____.
10. A heat converter consists of a/an _____ and _____.
11. A massive material within a solar collector converts sunlight into _____.
12. The two kinds of solar heating systems are _____ and _____.
13. What kind of air conditioner is best to cool an entire house?
14. Describe each of the four kinds of distribution systems.
15. Why would you ventilate a structure?
16. What purpose does an air exchanger serve?

ACTIVITIES

1. Try to get some plans showing how your school is built. How is it heated, cooled, and ventilated? Is the air filtered? Ask permission to inspect heating and cooling equipment areas.
2. Describe the energy source, converter, distribution system, and controls.
3. Visit a large city building. Prepare a report or display on how the climate is controlled.
4. Do you or your neighbors have solar collectors on your house? Ask to look at the piping and control system. Make a sketch of the system showing the valve, pump, collector, and tank locations.

If houses are well insulated when constructed, heating and cooling expenses will be lowered. (Pella/Rolscreen Co.)

Chapter 21

INSTALLING A PLUMBING SYSTEM

After studying this chapter, you will be able to:
☐ Identify plumbing fittings.
☐ Compare the jobs of a plumber and a pipefitter.
☐ Describe potable water systems, firefighting systems, and wastewater systems.
☐ Describe chemical piping systems.
☐ Explain how to join copper, plastic, steel, and cast iron pipe.
☐ Tell how to support piping and fittings.

Piping is used to move liquids and gases. Pipes transport liquids and gases from the supplier to the user. In the case of sewers, pipelines transport liquid waste away from the user.

The piping system inside the structure is called plumbing. It transports liquids and gases throughout the structure. The word "pipeline" is used only for the outside supply piping. Water can be supplied from either a private well or from a utility.

When water is supplied by a utility, the quantity used is measured with a *water meter.* The piping which follows the meter becomes smaller the farther it is from the meter. For example, often a 1 1/2 in. steel pipe follows a meter, and sizes decrease to 1/2 in. tubing.

A second major system handles waste. The common term for waste piping is DWV. The letters refer to "drain, waste, and vent." Pipes are cast iron, copper, ABS (acrylonitrile butadiene styrene), or PVC (polyvinyl chloride). Inside diameters (ID's) are 2, 3, 4, 6, 8, 10, 12, and 15 in.

DESIGNING PLUMBING SYSTEMS

The architect chooses the number, type, quality, and placement of *fixtures.* She or he also selects equipment and appliances that you do not see.

Pumps and water heaters are two examples of items that are not called fixtures.

Architects consider building codes. They find out how the building is to be used. The building is then built to meet that need.

Engineers design the piping systems that serve the fixtures. They specify the kind and sizes of pipes. They state where they go and how they are joined and attached. Some common DWV fitting types are shown in Fig. 21-1.

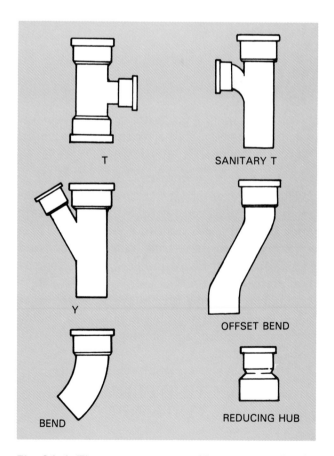

Fig. 21-1. These names are used for common plumbing fittings.

PLUMBING INSTALLERS

Plumbing systems are installed by subcontractors. *Plumbing subcontractors* do only plumbing. *Mechanical subcontractors* may do heating, cooling, and ventilating as well as plumbing.

Two groups of tradespeople who do the work are called plumbers and pipefitters. *Plumbers* install water, gas, and DWV systems. The systems are put into homes and commercial buildings.

Plumbers are on the site more than once. They place water supply and DWV lines before the basement floor is poured. They come again to rough in pipes in the walls before the walls are finished.

The plumbers return once more after the walls are enclosed to place and connect the fixtures. See Fig. 21-2.

Fig. 21-3. Pipefitters installed the pipes on this section of an oil refinery. (American Petroleum Institute)

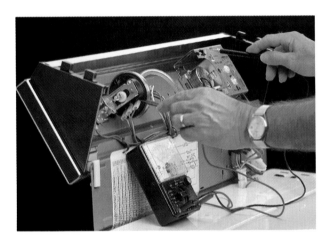

Fig. 21-2. Water flow in washers is sometimes controlled by electrical valves. A plumber may need some knowledge of electricity. (Whirlpool Corp.)

The second group of workers consists of the pipefitters. *Pipefitters* work with high pressure pipes and fittings in industrial structures. These carry hot water, industrial liquids, steam, and gases. Much of the work in Fig. 21-3 was done by pipefitters.

Plumbers and pipefitters may specialize, which means working in a special field. Hot water heating systems or residential plumbing are two special fields.

INSPECTING PLUMBING SYSTEMS

Plumbing systems are inspected by engineers and authorities. They see that the plumbing is done right. See Fig. 21-4.

Plumbing lines are checked for leaks, cleanliness, and flow rates. Compressed air is used to check for leaks in water and gas lines. A pressure gauge shows if the line is losing pressure. Smoke or a smelly gas is also used to find leaks. The leak will be near the smoke or smell.

Flow tests are run to see if the pipes meet specifications. A blockage or bend in a line can reduce the flow.

The Department of Health may check the cleanliness of the potable water system. They check for dirt, oil, and other deposits. Plumbers may need to sterilize the pipes.

Fig. 21-4. Every system must be carefully checked. Complex piping systems have many places where leaks can occur. (Standard Oil Co. of California)

PLUMBING SYSTEMS

Plumbing is used to:
- Provide fresh hot and cold water.
- Remove wastewater.
- Get water to fight fires.
- Heat and cool buildings with water and steam.
- Provide gases to burn and for industrial work.
- Provide fluids for industrial work.

This chapter discusses these systems: potable water, wastewater, firefighting, gas pipe, and process fluid systems.

Potable Water Systems

Potable (pronounced like "notable") *water* is pure enough to drink. In cities, water is made pure in a water treatment plant. Pipelines transport it throughout the city. The pipes are called *water mains.* They are made from precast concrete, clay, or cast iron. The mains are buried under or near the street.

A special connection is used to tap into a water main. The water company taps into the main and installs a *curb valve.* A water line runs from the main to the building. The shutoff valve and meter are placed inside. There they will not freeze.

Cold water pipes are run directly to the fixtures. See Fig. 21-5. A branch from the cold water pipe goes to the water heater. The *water heater* may be a part of the boiler or furnace which is used to heat the structure. See Fig. 21-6. Most home water heaters are tank type heaters. Tankless water heaters are

Fig. 21-6. The same energy conversion unit can be used to heat space and water. It saves energy. (Amana Refrigeration, Inc.)

being used more and more. See Fig. 21-7. Tankless heaters save energy.

Hot water lines should be insulated. This will reduce heat loss. Hot water lines run parallel to cold water lines. Hot water lines are on the left of a fixture. Look again at Fig. 21-5. Water lines go to each fixture that uses hot or cold water. Not all fixtures use both hot and cold water. Can you find any in Fig. 21-5?

Shutoff valves are used to turn off sections of the system. Shutoff valves should be placed in each line

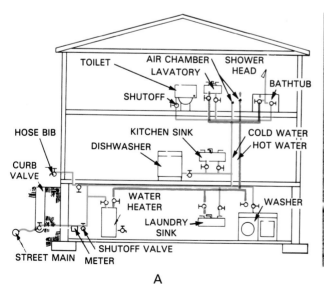

Fig. 21-5. Potable water systems. A–System with cold and hot water lines. Trace each line in this drawing. B–Sewer lines and drains are placed before the basement floor is poured. (Mary Robey)

Fig. 21-7. Tankless water heaters are small. They provide a steady flow of hot water. They do not store hot water. It is heated as you use it. (Controlled Energy Corp.)

at each fixture. It makes it easy to do repair work. Only the water to the fixture is turned off.

Air chambers absorb the shock of turning water off. Without them, pipes may bang when a faucet is turned off.

Appliances, fixtures, and equipment are attached to the plumbing system. Dishwashers and washing machines are *appliances*. Tubs, sinks, and toilets are *fixtures. Equipment* includes water heaters and water softeners.

Wastewater Systems

There are two kinds of wastewater in a community: runoff and sewage. *Runoff water* comes from rain and melting snow. This water runs off roofs, land, and paving, and runs into drains that lead to storm sewers. Storm water has less solid waste in it than sewage. It is not treated. *Storm sewers* dump the runoff into streams.

Sewage contains human and industrial waste. It goes into *sanitary sewers* that take it to a wastewater treatment plant. The waste products are removed

from the water and the clean water is returned to waterways. The *sludge* (remaining solid) is treated. The treated sludge is disposed of by burning, burying, spreading, or processing into fertilizer.

Private treatment systems, or *septic systems*, are sometimes used. In rural areas and many urban areas, the simplest or least costly disposal system relies on natural bacterial action in an underground tank. The tank is installed by a subcontractor who knows the local building code. An outdoor *leach* (absorption) *area* is usually provided. Plastic or cast iron piping and fittings are installed much like waste systems which connect to utilities.

Drain systems in buildings have traps, lines, vents, and cleanouts. The assembly is called a *DWV* (drain, waste, vent) system. Its parts are shown in Fig. 21-8. The *trap* keeps sewer gas from escaping into the structure. Sewer gas is formed by decaying waste materials. It smells bad and is dangerous to people.

DWV lines are made of cast iron, copper, or plastic. The branches of DWV lines all connect to the house sewer. The lines near the sewer are larger than those at fixtures.

DWV lines need vents. Refer to Fig. 21-8 again. A *vent* lets sewer gas escape to the outside air. Water in traps is siphoned out if drains are not vented.

Wastewater pipes all run downward. They drop at least 1/4 in. for each 1 foot of length. Gravity moves the wastewater along. The pipes plug up easily.

Cleanouts are used to remove a blockage. A cleanout cap or plug is removed and an auger or snake is put in. These devices either pull out or cut up a blocking object.

Toxic waste

Toxic wastes cannot be treated in water treatment plants. *Toxic wastes* are wastes that cause health problems and perhaps death. These wastes are handled in a separate plumbing system. They are piped to a place where one of the following is done. They are:

- Put into containers.
- Permanently stored.
- Burned.
- Processed.

Firefighting System

The *firefighting system* is separate from the potable water system. There are two kinds. See Fig. 21-9. The standpipe and sprinkler systems are most common.

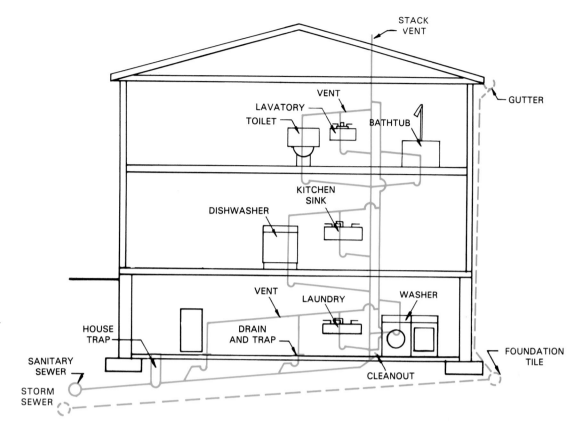

Fig. 21-8. Drains are sloped. Gravity makes the water flow. Parts include traps, vents, and cleanouts.

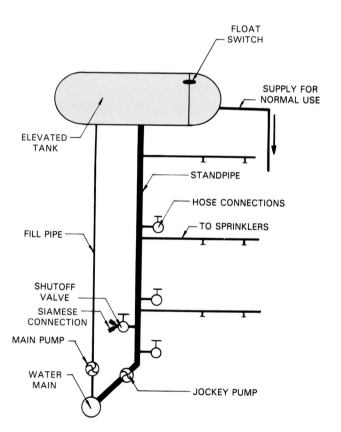

Fig. 21-9. A secure supply of water is needed in a firefighting system. Can you point out three ways a supply of extra water is assured?

A *standpipe system* has a permanent source of water and hose connections. Each floor or section of the building is served.

The permanent source of water may be a rooftop tank. Water for normal use is drawn from the upper part of the tank. That way, there is always a supply to fight fires. A float switch in the tank checks the water level to keep it full. As water is used, the main pump is turned on to refill the tank.

A pump called a *jockey pump* maintains water pressure in the system. It works on a pressure switch. When the standpipe pressure falls, the pump restores it. The jockey pump is connected to the city water supply.

A unit called a *siamese connection* is on the outside of the building. See Fig. 21-10. It has two caps on it. Two pumper fire trucks with two 2 1/2 in. lines can be joined to the connector. Two pumper lines are used to feed a 4 - 6 in. standpipe line.

The *sprinkler system* is automatic. A sprinkler nozzle holds back the water. When a space gets hot enough, a plug melts and the water begins to flow. At the same time, the jockey pump starts. The system sounds an alarm throughout the building and at the fire station. This system provides a 24 hour per day firewatch.

Fig. 21-10. A fire truck called a pumper can supply water through a siamese head.

Gases for Burning or Doing Industrial Work

Plumbing systems may transport gases. The gas may be for heating or doing other work. Natural gas is piped to furnaces, water heaters, clothes dryers, and cook stoves. Compressed air is used to run tools and inflate tires.

Gas lines have a cleaning drop, lines, and shut-offs. The *cleaning drop* cleans the gas. See Fig. 21-11. Dirt in the line settles into the drop. The lines are steel pipes. Large lines are welded or use flanges. Smaller pipes have threaded fittings.

Compressed air systems have more parts. See Fig. 21-12. Parts are a compressor, tank, air line, filter, regulator, pressure gauge and switch, relief and shut-

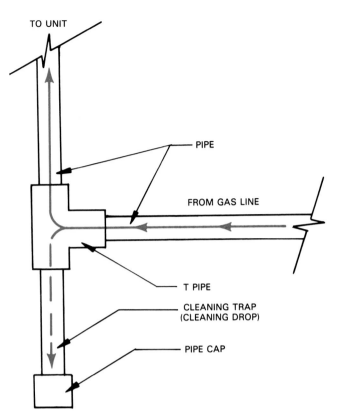

Fig. 21-11. Gravity causes dirt to fall into the cleaning drop. Dirt can clog burners if it is not removed.

off valves, and connectors. Compressors raise the pressure of the air in a closed system. The tank stores the air. The compressed air is moved through steel or copper lines. Filters remove oil, water, and dirt from the air. The air pressure is adjusted with a regulator and gauges show the pressure in the system. Flexible hoses for air tools are attached to the connectors in

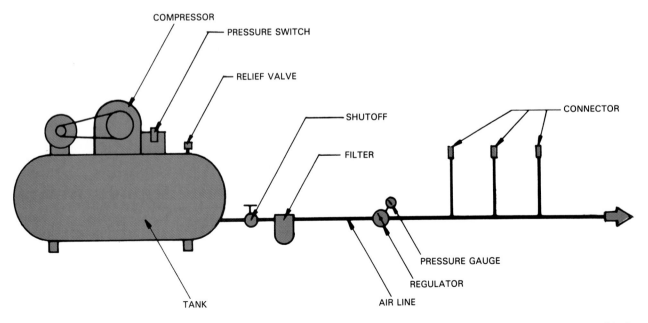

Fig. 21-12. Compressed air systems must be well sealed. Air at high pressures can leak from a very small hole.

the lines. The relief valve and pressure switch maintain a safe pressure. Shutoffs are used in the same way as in other systems.

Process Fluids

Process fluids vary greatly. Cooling water, chemicals, and oils are *process fluids*. Some corrode metal. Others must be kept at high pressures. Special systems are designed to move them, with each system kept separate. The system design is based on three things: liquid quantity, temperature, and pressure.

INSTALLING PLUMBING SYSTEMS

Plumbers cut pipe and tubing and join the pipe with fittings. The piping assemblies are attached to walls, floors, and ceilings. The piping is joined to supply lines. This is called roughing-in. See Fig. 21-13. Most people do not see this work. It is covered when the inside wall coverings are installed.

Copper Tubing

Copper tubing is used for water lines. It may be for hot or cold, clean or dirty water. Larger copper pipes are used for drain pipes. Copper does not corrode as badly as steel pipe. It is easy to cut and join. One kind of copper is soft and bends easily. The other kind is rigid. It is not often bent.

Copper or brass fittings are used to join pieces. See Fig. 21-14. The fittings slip over the ends of the

Fig. 21-13. Rough-in work will be covered after it is finished.

pipe. Solder is used to seal one kind of joint. *Solder* is a mixture of lead and tin. The pipe and fitting are heated. The solder melts and runs into the joint. This type of connection is known as a *sweat joint*.

Copper tubing is also joined by mechanical methods. The *flare* and *compression joints* are the common methods used. See Fig. 21-14.

Plastic Pipe

The cost of plastic pipe is low. You can buy several kinds. Some are stiff, others you can bend. All

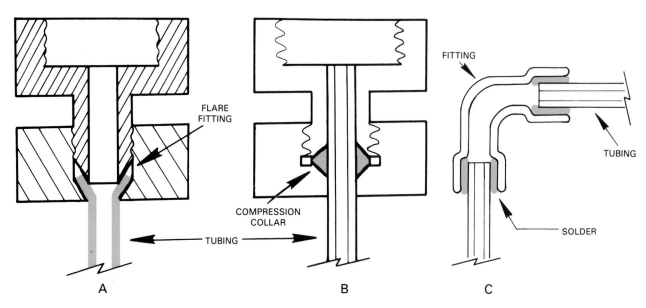

Fig. 21-14. Copper tubing is joined with flare, A, compression, B, and sweat, C, joints. Mechanical (compression), B, and cemented joints are used with plastic pipe.

plastic pipe can be used for cold water pipes. Only certain kinds are used with hot water. The large sizes work well for drains.

Each kind has its own fittings and method for joining. Some joints are similar to sweat joints. In plastic, they are called *cemented joints*. The pipe is cut to length and the burrs on the end are removed. A cement is spread to cover the outside end of the pipe and the inside of the fitting. The pipe is inserted into the fitting. In a short time the joint dries and becomes strong.

Mechanical joints are used with certain kinds of plastic pipe. *Compression fittings* are much like the copper fittings that were shown in Fig. 21-14. Clamps are used to seal the joint on others. Clamp joints are not used for high pressures.

Steel Pipe

Steel pipe is very strong. However, chemicals in water combine and form a weak acid. In time, the acid can corrode the pipes. Minerals in the water often collect inside the pipes. They form a *scale* which builds up and makes the inside of the pipe smaller.

Steel pipe less than 3 in. in diameter is joined with *threaded fittings*. Pipe lengths over 8 in. are cut and threaded on the site. The shorter lengths are called *nipples*. They are bought as standard items already threaded. Fittings are sold with threads already on the inside.

Pipe threads are tapered. The more you tighten them the tighter they seal. Special plastic (Teflon©) tape and pipe dope help to seal the joint. Pipe dope is a thick liquid. It fills any gaps left between the threads.

Large diameter pipes are not often threaded. Large pipes are welded or held with flanges. See Fig. 21-15. Gaskets are used to seal *flange joints*. A *gasket* is a soft, compressible material that is squeezed at the joint. *Welded joints* are hard to take apart. Most of the joints in Fig. 21-16 are welded.

Cast Iron Pipe

Cast iron pipe is often called soil pipe. It is usually used for sanitary drains and sewers. The two ends of soil pipe differ. The small end is called the *spigot*. The larger end is the *bell*. The spigot end is placed in the bell end. The joint is sealed in three ways. Rubber gaskets, oakum and lead, or a sleeve and clamps are used.

Fig. 21-15. Flanges and welds are used to join these pipes. Flange joints can be taken apart. (Amoco Corp.)

Fig. 21-16. Welded joints are permanent. The workers must be very careful. Can you find any welded joints in this picture? (Shell Oil Co.)

SUPPORTING PIPES

Pipes are fastened to the structure. When water is turned off and on, the pipes move. Constant movement can crack the joints, causing leaks.

Pipes that run horizontally are supported with *hangers.* Hangers are attached to framing in the floors and ceilings. *Backing boards* are added to the framework. Metal and plastic clamps are used to secure vertical pipes.

SUMMARY

Piping systems move liquids and gases. Pipelines run between structures, while plumbing systems are inside the structures. Engineers and architects design piping systems. Plumbers and pipefitters install the system. They work for subcontractors. They run lines to the site, rough-in piping, and install fixtures.

Plumbing systems consist of piping, processing equipment, appliances, and fixtures. Each plumbing system is kept separate.

Piping consists of pipes and fittings. Parts are made of copper, plastic, steel, and iron. Each is joined in a special way. Piping is held in place with hangers, clamps, and special framing.

KEY WORDS

All of the following words have been used in this chapter. Do you know their meaning?

Air chamber
Backing board
Bell
Cemented joint
Cleaning drop
Cleanout
Compression fitting
Compression joint
Curb valve
DWV
Firefighting system
Fixture
Flange joint
Flare joint
Flow test
Gasket
Hanger
Jockey pump
Leach area
Mechanical subcontractor
Nipple
Pipefitter
Plumber
Plumbing subcontractor
Potable water
Process fluid
Runoff water
Sanitary sewer
Scale
Septic system
Sewage
Shutoff valve
Siamese connection
Sludge
Solder
Spigot
Sprinkler system
Standpipe system
Storm sewer
Sweat joint
Toxic waste
Trap
Treaded fitting
Vent
Water heater
Water main
Water meter
Welded joint

TEST YOUR KNOWLEDGE

Write your answers on a separate sheet of paper. Do not write in this book.

1. A firm that installs plumbing systems is most often a _____.
2. People who install water, gas, and sewage systems are (pipefitters/plumbers).
3. Pipefitters install _____ pressure pipes.
4. Piping systems are inspected for _____, _____, and _____.
5. Water pure enough to drink is called _____ water.
6. Pipes that run directly to fixtures carry _____ water.
7. Pipes will not bang if they have _____ _____ on the lines.
8. The two kinds of wastewater are _____ and _____.
9. Sewer gas is kept from entering a structure with a/an _____.
10. Water is moved through sewer pipes by _____.

11. There are two kinds of firefighting systems. They are _____ and _____.
12. A compressor, tank, lines, filter, regulator, and gauges are used in a/an _____ _____ system.
13. Three common materials used for water lines are _____, _____, and _____.
14. Plastic pipes are joined with _____ and _____ joints.
15. Steel pipe is joined with _____ fittings.
16. The larger end of a cast iron pipe is called a spigot. True or false?

ACTIVITIES

1. Talk to plumbers about their work. Find out what the work is like, what tools are used, and what the working conditions are.
2. Inspect your house plumbing. Find the shutoff valves and water heater. What kind of piping was used?
3. Talk with a custodian at school. Ask to see the mechanical room. Have him or her point out and explain the various pieces of plumbing equipment.

Chapter 22

ROUGHING-IN ELECTRICAL POWER SYSTEMS

After studying this chapter, you will be able to:
- ☐ Describe the correct order for installing a power system in a structure.
- ☐ Describe the purpose and operation of a service panel box.
- ☐ Explain what a branch circuit is.
- ☐ List types of cable, junction boxes, switches, tubing, and receptacles.
- ☐ Describe ways to plan circuits for concrete construction projects.
- ☐ Use color-coding to identify wires.

Electrical systems distribute electricity throughout the country and within structures. There are two kinds of electrical systems. One supplies power, and the other is a communication system.

The *power system* supplies the electrical energy needed to run the mechanical equipment in the building. See Fig. 22-1. Power is used for heating and cooling, and lighting.

Fig. 22-1. These large wires supply power to an apartment building. Each circle is a place for attaching a meter to measure electricity used. There is a meter for each apartment. (General Electric Co.)

Electricity also powers the appliances and machines used by the people who live and work in buildings. Families use electric stoves, dryers, refrigerators, and sound equipment. In factories, most machines that produce products are powered by electric motors.

Electrical *communication systems* transmit information with electrical impulses. They use lower voltages and smaller wire. These systems are covered in a later chapter.

ELECTRICAL POWER SUBSYSTEMS

The electrical power system has many smaller systems. These include the power source, distribution lines, service cables, service entrance, and branch circuits.

Source

Electricity is made in a *power generating plant.* Water, coal, oil, nuclear materials, and other energy sources are used to power the generators. See Fig. 22-2.

Distribution

Electricity is sent to *distribution stations.* It is sent at very high voltages. See Figs. 22-3 and 22-4.

Cables carry the power to local areas. Transformers lower the voltage and service lines bring the power to individual buildings. Outside the structure is a *service transformer.* See Fig. 22-5. It lowers the voltage to that needed in the structure.

Service Cable

Power is carried from the service transformer to the structure through a heavy insulated wire called

Fig. 22-2. Electric power is produced at this nuclear plant. (New York State Power Authority)

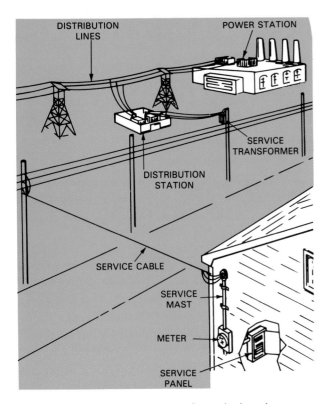

Fig. 22-3. Simple diagram of total electric power system. Because electricity leaves the power plant at high voltage, the distribution station and small transformers reduce the voltage so it can be used in buildings. (General Electric Co., Wiring Devices Dept.)

Fig. 22-4. These transformers lower the voltage for distribution to customers.

Fig. 22-5. This worker is installing a new step-down transformer. High-voltage electricity is carried by the lines. A step-down transformer lowers the voltage. (Commonwealth Edison Co.)

a *service cable.* It may either be buried or installed overhead. The overhead cables enter through a *service mast.* This is a pipe or conduit with a cap on it to keep out water. See Fig. 22-6. The size of wires in the service cable depends on the amount of power needed. The more electric power, the thicker the cable. Published tables help select the right size.

The service cable is wired to the meter base. This is where the meter is attached. The *meter* measures power used. It is placed outside the structure where it is easy to read.

The power company installs the meter after the building electrical wiring is installed. The work must

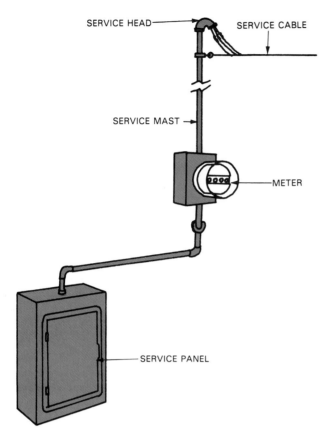

Fig. 22-6. The power company installs the service cable. They sometimes install the meter base. The owner does the rest of the work.

pass inspection before the meter is installed. The power company's responsibility ends at the meter.

Service Entrance

A *service entrance* connects the meter with the service panel. Fig. 22-6 shows the service entrance. The service panel is installed out of the way, but made easy to reach. See Fig. 22-7. The *service panel* consists of a steel box, switches, and circuit breakers. The steel box is strong and fireproof. All wires are attached within it. The box houses the main breaker. The *main breaker* is a switch that shuts off all power to the building. It should be turned off before working in the service panel.

Branch Circuits

A *branch circuit* is all of the wiring, electrical outlets, and lights controlled by one fuse or circuit breaker. If you look at Fig. 22-8 you can see three different branch circuits. All of them start out inside the service panel. Some circuits, such as the one to the washing machine, supply electricity to only one appliance or outlet. Others, such as the one going to the lights, may have other things connected to them. This is done to balance the amount of electric

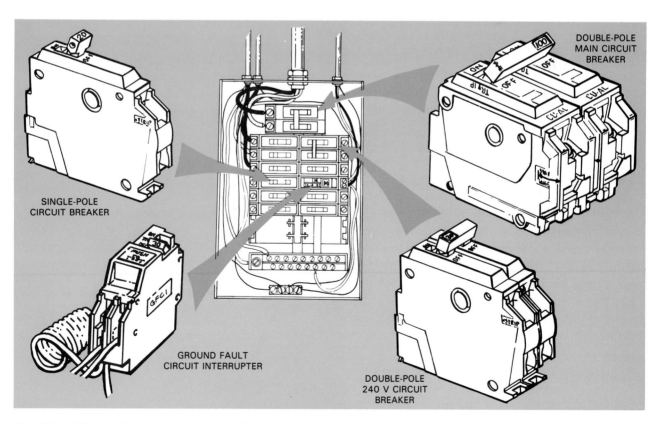

Fig. 22-7. The main circuit breaker and branch circuit breakers are in the service panel. The main breaker is at the top. At center and below are single pole, double pole, and GFCI types. (General Electric Co.)

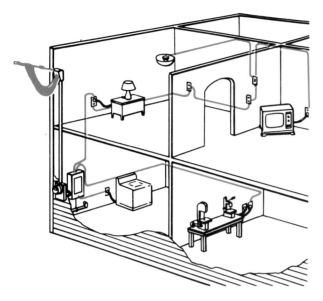

Fig. 22-8. How to define a branch circuit. One branch circuit in the upper left room supplies the table lamp, rear wall outlets, ceiling light, and ceiling light switch. A second branch circuit supplies the two workshop machines. Each branch circuit begins at a separate circuit breaker.
(General Electric Co., Wiring Devices Department)

current in each circuit. Since appliances require more electric power than lights, one appliance may be all that the circuit can safely carry.

A fuse or a circuit breaker inside the service panel makes the final connection between the branch circuit and the electric power coming into the panel.

There are general purpose, small appliance, and fixed appliance circuits. *General purpose circuits* are used for lights and outlets. These circuits are not used in kitchens. Kitchens have *small appliance circuits.* They supply toasters, cookers, and mixers. These circuits often have special circuit breakers and carry larger loads.

Each large piece of equipment has its own circuit. It is called a fixed appliance circuit. Clothes dryers and electric ranges are wired this way. Only one appliance is on a *fixed appliance circuit,* whereas one or more can be on a small appliance circuit. In industrial buildings, circuits used for lighting and machines are kept separate. Each large machine has its own circuit. Production lines often have very complex circuits. Fig. 22-9 shows a large control panel.

INSTALLATION REQUIREMENTS

The power system inside the structure has many parts. They are circuit breakers and fuses, conductors, supports, junction boxes, receptacles, controls,

Fig. 22-9. Complex controls play a big role in making glass. This console regulates the flow of special sand and other materials put in a melting furnace. (Libbey-Owens-Ford Co.)

and loads. Some knowledge of their operation is needed to make proper connections with the least amount of work.

Fuses and Circuit Breakers

Fuses and circuit breakers are in the service panel. Always remove the fuse or turn off the circuit breaker before working on a circuit. Fuses and circuit breakers keep branch conductors from overheating and causing a fire.

Fuses are shown in Fig. 22-10. When too much current is drawn through a *fuse,* a wire in the fuse melts (burns out).

There is more than one kind of circuit breaker, Fig. 22-11. *Circuit breakers* are preset to allow a certain amount of current to flow through them. When too much current flows, circuit breakers switch off the circuit.

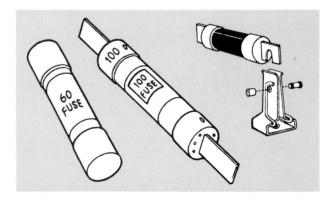

Fig. 22-10. A–Cartridge fuses. B–This Class R type fuse is commonly used with a safety switch. (Square D. Co.)

A

B

Fig. 22-11. Circuit breakers work like switches. They are large. A–This one carries 300 amps of current. B–How much will this breaker carry? (General Electric Co.)

Fig. 22-12. Ground fault circuit interrupters can protect one receptacle or an entire circuit from shock. The three front devices protect you from shock when using an unprotected circuit. (Leviton Manufacturing Co.)

Common circuit breaker types are switch and pushbutton. The *switch type* uses toggle handles. They look and work like wall light switches. Pushbutton circuit breakers are different. The *pushbutton* type has a button that extends out when an overload occurs. To reset it, just push the button.

Some circuit breakers have a slight delay before they turn off. When motors start up there is a surge of electricity. The delay allows for temporary overloads.

There are circuit breakers that protect against more than overloads. They protect people against a common type of electrical shock. It is called the "ground fault." Fig. 22-12 shows several *ground fault circuit interrupters (GFCI).* A ground fault occurs when a person serves as a conductor. A tool with a short can cause a ground fault shock. Ground fault circuit breakers trip in a fraction of a second. Most healthy people would not be hurt. This protection is required where outlets are near water. Double pole GFCI breakers are used in 240 volt circuits.

Conductors

Conductors carry electrical power. They come in the form of wires or cables. Copper is used the most. See Fig. 22-13. Aluminum does not conduct electricity as well as copper. It is sometimes used because it costs less and is lighter in weight.

Conductors are made of solid wire or stranded wire. Stranded wire is more flexible than solid wire of the same size. Large stranded wire (over #10 gauge) is called cable.

The current carrying capacity of wire depends on two things. The larger the diameter of the wire the more current it carries. The kind of metal used to make the wire affects the flow and the cost. Copper is the best reasonably priced metal.

Wire and cable are covered with *electrical insulation.* See Fig. 22-14. Insulation keeps wires and cables away from things they should not touch. A "live" wire has electricity flowing through it. If people touch live wires, they may get shocked. People get shocked when electricity flows through their bodies. Small shocks hurt. Large ones can kill.

Rubber and plastic are used to insulate wire. The type and thickness depends on the maximum voltage and kind of exposure, As voltage increases, thicker insulation is needed. This does not apply to

Fig. 22-13. Many kinds of cords and cables are used in distributing electrical power throughout a structure. (Leviton Manufacturing Co.)

some outdoor lines. Very high voltage lines may have no insulation. Instead, the lines are held high above the ground. People and most machines cannot reach them. You should not fly kites near these lines.

Special insulation is needed when cables are:
- Buried underground.
- Encased in concrete.
- Used in a corrosive air.

You can buy cable with more than one conductor in it. The most common have two, three, or four conductors in them. The insulation is color coded. This helps the electrician keep track of the wires. *White* is the **neutral.** No switches or fuses are placed in the white wire. *Green* or bare wire is used for safety. It is called the **ground.** *Black* and *red* wires are the **"hot" wires.** Switches and fuses are put into these lines. Other colors are used in complex wiring systems. They are placed in conduit and raceways. These are discussed later.

Two-conductor cables have a white and black wire. A red is added to three conductor cables. A green or bare wire can be added to two and three wire cables. Fig. 22-14 shows these cables. The size of wire and outside covering varies greatly.

Conductor Supports

Conductors must be supported and protected. Cables in residential construction get support from framing members. There are many ways to fasten cables to framing. Fig. 22-15 shows hardware used for fastening conductors.

Cables in public, commercial, and industrial structures require more protection and support. These circuits often use high voltage and receive a lot of vibration and wear. These wires are run through conduit and raceways.

Conduit consists of metal or plastic tubes. **Rigid conduit** is much like water pipe but is somewhat thinner. Threaded fittings are used to join it. All

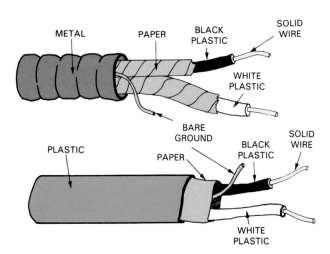

Fig. 22-14. Conductors are covered with insulation. The ground wire is left bare. Plastic and paper hold them inside of a cable.

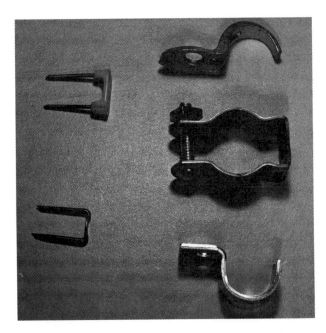

Fig. 22-15. Electric cables and conduit are held to the structure's frame with these devices.

pipes and bends are threaded on both ends. They are joined with *couplers.* That way, the inside of the conduit is kept smooth. When wires are pulled through it, they will not catch. Rigid conduit comes in sizes up to 4 in.

Circuits hidden inside wall, floor, or ceiling framework usually consist of metallic and nonmetallic cable. However, any exposed circuit sections must use conduit.

Thinwall conduit can be bent. Special tools are needed. See Fig. 22-16. Thinwall conduit is often called electrical metallic tubing (EMT). It comes in 1/2, 3/4, and 1 in. sizes. The tubing length is 10 ft. Fittings are used to join pieces together.

Plastic conduit can be used when wires go underground or in concrete. It does not corrode.

Raceways are made of formed metal or plastic, Fig. 22-17. They have snap-on covers and are

Fig. 22-16. Conduit protects conductors. It also provides an organized conductor path. (American Electric)

Fig. 22-17. Raceways are used when walls of a structure are solid. (American Electric)

mounted on the surface of walls when remodeling. They are used in dry places. Voltage is less than 330 volts

Raceways may be a part of the floor in new construction. They are a part of the floor decking. They may supply power to space above or below, as in Fig. 22-18.

Fig. 22-18. Floor decks have raceways placed in them. They save time. (American Electric)

Junction Boxes

Splices and connections between electrical conductors are made in *junction boxes.* Junction boxes usually have four or eight sides. A few are round. Fig. 22-19 shows common shapes. Some boxes are fiberglass. *Gang boxes* are used when two or more boxes are hooked together. They are used when more than one switch is at the same location.

Receptacles are used to quickly disconnect and connect equipment to a circuit. See Fig. 22-20. Different kinds are used for different voltages. Fig. 22-21 shows several locking devices.

Controls

Switches are the simplest power controls, Fig. 22-22. They open and close a circuit. A *"single-throw" switch* opens and closes one wire in a circuit. *Three-way switches* are used when a load (unit

A

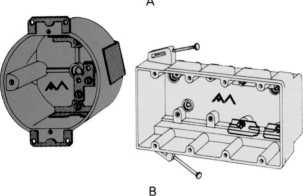

B

Fig. 22-19. Junction boxes vary in size, shape, materials, and design. Some can be ganged together. A–Metal units. B–Molded fiberglass units. (Allied Molded Products)

Fig. 22-20. There are many kinds of receptacles. Most are used with 120 volt connectors. Which one can be used with higher voltages? The gold one has a dust cover and is used in the floor. (Leviton Manufacturing Co.)

being powered) is controlled from two points. A load may be controlled from more than two points. In that case, *four-way switches* are needed.

Dimmer switches vary the voltage in a circuit. See Fig. 22-23. They raise and lower the brightness of an incandescent light.

Fig. 22-21. Locking devices keep plugs from pulling out of the receptacle. (Leviton Manufacturing Co.)

When you are gone, appliances can be turned on with a *timer.* Lights may be turned on so that it appears someone is home. Timers can even start breakfast while you are still sleeping.

Lights in office buildings use a lot of energy. They can account for over half of the energy consumed. New light controls save a large part of that energy. Both timers and dimmers are used. Timers turn lights off when no one is using the building.

A dimmer for a commercial lighting circuit works automatically. A *photohead* detects light levels. The controller automatically adjusts the dimmer. The control system maintains an even brightness. When the sun is bright, lights are turned down or off. When new lighting tubes brighten the area, the controller lowers the light level. When a government standard requires less light, the whole lighting system is turned down.

The devises in Fig. 22-24 make life easier for all users. Remote control, motion sensors, lighted devices, and sound sensors make the controls easier to use.

Electrical Loads

Lights, motors, heaters, and electronic equipment are *loads.* All homes have many loads. The number and size of loads determines the size of the service entrance. The load is measured in the number of amperes (amps) it draws. Electrical appliances have nameplates that state the number of amperes they draw. Most home circuits can provide

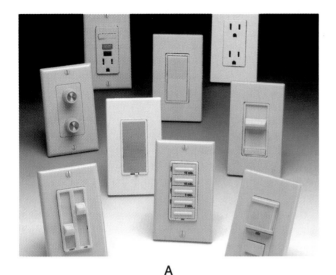

A

B

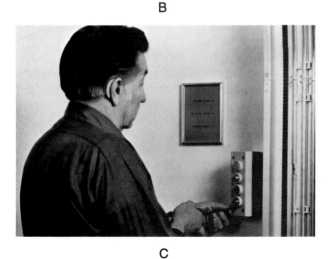

C

Fig. 22-22. Controls. A—Receptacles, switches, dimmers, fan speed controls, and motion detectors. B—Remote, programmable, and manual controls for home lights and appliances. Existing AC wiring can be used. (Leviton Manufacturing Co.) C—Heavy-duty switches control large equipment. (TransLogic Corp.)

Fig. 22-23. Many kinds of manual and automatic devices control the brightness of lights. (Leviton Manufacturing Co.)

Fig. 22-24. These specially designed devices are easier to use than some switches and controls commonly found in homes. (Leviton Manufacturing Co.)

15-20 amperes of current. A circuit is overloaded when the total load is higher than the planned current.

INSTALLING AN ELECTRICAL POWER SYSTEM

The electrical power system includes the source, service entrance, and branch circuit. The power company is responsible for everything down to the meter. The subcontractor installs the service panel and branch circuits.

Installing Service Panels

Service cables run from the meter base to the service panel to bring in the power. Fig. 22-25 shows a heavy-duty service panel. A circuit breaker is installed for each circuit.

Fig. 22-25. A large switch is being installed in a service panel. (General Electric Co.)

Knock-out holes are partly punched. They are in the top and bottom of the service box. Cables feed through these holes. Wire clamps or conduit anchors mount through the holes and hold the cable or tubing.

Roughing-in Branch Circuits

The term "roughing-in" means to install a circuit, but not to install receptacles, switches, or light fixtures. The first step is to place junction boxes. They are fastened to structural members.

In concrete construction, junction boxes are fastened to the insides of forms. See Fig. 22-26. Conduit is run between all boxes. It provides a route for the wire.

After the forms are removed, wires are pulled through the conduit. In wood frame homes, cables are run through the frames. Cables pass through holes bored in the framing members. The holes are drilled near the center of the studs or boards. Drilling near the center weakens them the least. There is also less of a chance for nails to hit the cables. Sometimes holes must be drilled close to the framing edges. Steel plates are fastened to the edges to protect the cable from screws and nails.

Fig. 22-26. The junction boxes have been laid loosely at planned locations. They will be joined with conduit. (Wire Reinforcement Institute)

Cables are secured to the frames with special devices. Look again at Fig. 22-15. Connectors are used to attach the cables to the junction box knock-outs.

When conduit is used, fish tape is used to pull wires through the conduit. *Fish tapes* are long flexible metal strips. They are pushed through conduit. Wires are tied to the ends. They are then pulled through the conduit. These wires are not immediately connected to switches or circuit breakers. They are left coiled in the junction boxes. Switches and other devices are installed after the frame is enclosed.

Although the roughing-in step does not include connecting the receptacles, switches, or fixtures, you must begin with a sketch showing how many units are used and how each terminal is connected. This information allows you to count how many conductors you need in each cable or conduit section and helps you plan the wire colors needed. Figs. 22-27 and 22-28 show how to plan receptacle wiring and switch circuits.

For a receptacle, attach the white wire (neutral) to the silver colored terminal and the black or red wire (hot) to the brass colored terminal. A bare wire (ground) is attached to both a green screw on the receptacle and to any screw available on the inside of the junction box.

For a switch, the white wire must always be left continuous. All white wires within one box must be connected together. Attach the switch terminals only to black or red wires. If a white wire must be attached to a switch screw, paint the wire black or red. The problem arises because the cable from a light fixture may have a white wire in it, even though that wire acts as a "hot" wire.

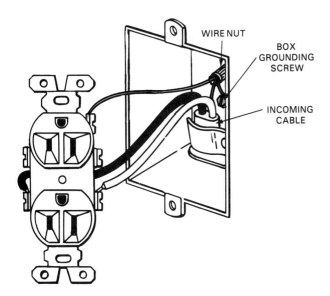

Fig. 22-27. Connect the white wire to the silver colored terminal on the receptacle. Put black wire on brass colored terminal. Make ground (green or bare) wire connections between receptacle, box grounding screw, and cable. A sketch allows you to count the wires needed during rough-in.
(General Electric Co., Wiring Devices Department)

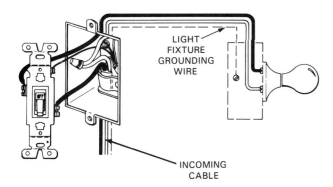

Fig. 22-28. Single pole switch wiring.
(General Electric Co., Wiring Devices Department)

SUMMARY

Electricity is made at a power generating plant. A transmission system distributes the power. Power comes to a building through a service cable. The power company's responsibility ends at the meter base. Subcontractors install the service entrance. Branch circuits are roughed in. The local electrical inspector checks the work. Then the frames are enclosed. Engineers and government people inspect electrical work in nonresidential buildings.

KEY WORDS

All of the following words have been used in this chapter. Do you know their meaning?

Branch circuit
Circuit breaker
Communication system
Conductor
Conduit
Coupler
Dimmer switch
Distribution station
Electrical insulation
Fish tape
Fixed appliance circuit
Four-way switch
Fuse
Gang box
General purpose circuit
Ground
Ground fault circuit interrupter (GFCI)
Hot wire
Junction box
Knock-out hole
Load
Main breaker
Meter
Neutral
Photohead
Plastic conduit
Power generating plant
Power system
Pushbutton
Raceway
Receptacle
Rigid conduit
Service cable
Service entrance
Service mast
Service panel
Service transformer
Single-throw switch
Small appliance circuit
Switch type
Thinwall conduit
Three-way switch
Timer

TEST YOUR KNOWLEDGE

Write your answers on a separate sheet of paper. Do not write in this book.

1. List and describe the five subsystems of an electrical power system. Start at the generating plant and end at branch circuits in a structure.
2. Circuits are made up of eight parts. Describe the function of each.

3. The first job when roughing-in a circuit is to place the _____ _____.

4. What is the last job in roughing-in a branch circuit?

Matching questions: On a separate sheet of paper, match the definition in the left-hand column with the correct term in the right-hand column.

5. _____ The device protects against electrical shock.
6. _____ Color of neutral wire.
7. _____ Color of "hot" wire.
8. _____ Makes joint between conduit lengths.
9. _____ Pulls wire through conduit.
10. _____ The device protects against overloads.

a. Red or black.
b. Circuit breaker.
c. Ground fault circuit interrupter.
d. White.
e. Coupler.
f. Fish tape.

ACTIVITIES

1. Find the service line to your home or school. Does it have a two or three wire system?

2. Locate the circuit breakers or fuses. How many circuits does your home have? What is the ampere rating of each? Find out what is on each circuit.

3. Make a list of materials used in the branch circuit. List only those you can see without taking anything apart.

4. Plan three branch circuits: for lamps, portable kitchen appliances, and a clothes dryer (30 amp). Where will the wires go? What kinds of cable will be used? List all parts to be used.

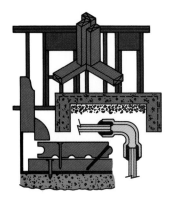

Chapter 23

INSTALLING COMMUNICATIONS SYSTEMS

After studying this chapter, you will be able to:
☐ Describe ways to transmit sound, video, and sensor information.
☐ Recall wiring color codes for telephone circuits.
☐ Design a doorbell circuit.
☐ Choose monitoring systems or those that provide a two-way exchange of data.

Homes, schools, and office buildings must have a system for communicating. To inform each other and to extend our senses and our control over events, we use equipment.

KINDS OF SYSTEMS

The equipment may be simple or complex. Doorbells are an example. Some apartments and houses have an intercom and electric door locks, Fig. 23-1 and Fig. 23-2. Signs are used to give directions and warnings.

Fig. 23-1. An intercom system makes an apartment safer. The system uses one loudspeaker for both a microphone and a sound producer. (NuTone)

Fig. 23-2. An intercom has a sender/receiver in an apartment. Wire between the stations is 24 to 28 gauge in diameter. (NuTone)

Telephone, television, and computer networks help us communicate with the world and gather information.

A communication system sends and receives information. There are two groups of systems: monitoring and exchange systems.

MONITORING SYSTEMS

Monitoring systems watch a place, a machine, or a process. Television systems are set up to watch places both inside and outside. A person can watch an entire factory complex with television monitors, Fig. 23-3.

The performance of machines is communicated to control rooms. Sensors on the machine detect how the machine is working. The information is sent to a control room. A person receives the information and decides what to do. Fig. 23-4 shows a control room at a power plant.

Fig. 23-3. Television cameras are electronic eyes. They watch and transmit what they see to a monitor. One person can watch an entire factory from one place.

Fig. 23-4. The operation of the entire power plant is monitored and controlled from this room. The men are using a computer. It retrieves information and helps make decisions. (Southern California Edison Co.)

The simplest sensor is a switch operated by some motion of the machine being monitored. This switch is connected by wires to the control room panel. The operator can note the machine position on a panel light.

Sensors also monitor a process. The quality of water is checked with instruments at a water treatment plant. Instruments check for visual clarity and chlorine content. Water quality is recorded on charts. As an example, the visual clarity can be checked with a photocell (a light level "eye") sensor connected to an electronic chart plotter some distance away.

EXCHANGE SYSTEMS

Exchange systems involve people or equipment giving and getting information. Telephones, television, intercom, and computer systems are of the exchange type.

Engineers design communication systems. Subcontractors install the systems. Each type of system has its own unique connectors. See Fig. 23-5.

Telephone equipment can transmit a person's voice. See Fig. 23-6. We can talk with people thousands of miles away.

Telephone equipment transmits voice and pictures. Events can be watched from across the room or from the moon.

Information and records can be stored and retrieved with computers. We can learn, solve problems, and be entertained by information gotten with computers. Fig. 23-7 shows a person using a computer to make a drawing.

All of these communication systems need to be installed. Some are installed by the firm that offers the service. Telephone service and television cable are two examples. Others are installed by the owner or contractors.

Fig. 23-5. These connectors are used in communication systems. It is a sampling of the hundreds of different connectors made by one company. (Leviton Manufacturing Co.)

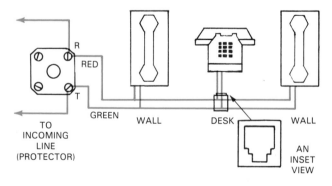

Fig. 23-6. A typical telephone circuit. Note wire colors to be connected.

Fig. 23-7. Computers are used for making drawings. This student is learning how to draw with computer graphics using a terminal. The main computer is a city block away. It could be hundreds of miles away. (Dover Elevators)

Telephone Systems

With a *telephone system,* it is easy to talk to almost anyone in the world. A large complex system serves mobile users. It consists of telephones, radio links, and exchanges. See Fig. 23-8.

Telephones come in many forms, Fig. 23-9. Telephones send and receive messages with electricity and light. The messages travel through wires, glass fibers, and space.

To send a message, telephones convert a person's voice into a modulated (varying) current. The current travels over wires. To receive a message, a modulated current is changed back to sound waves we can hear. Many newer systems send messages digitally (in discrete steps.)

Telephone wires are called *cables* when several wires are run together. They are not often seen. The cables are placed in walls, above ceilings, and under floors.

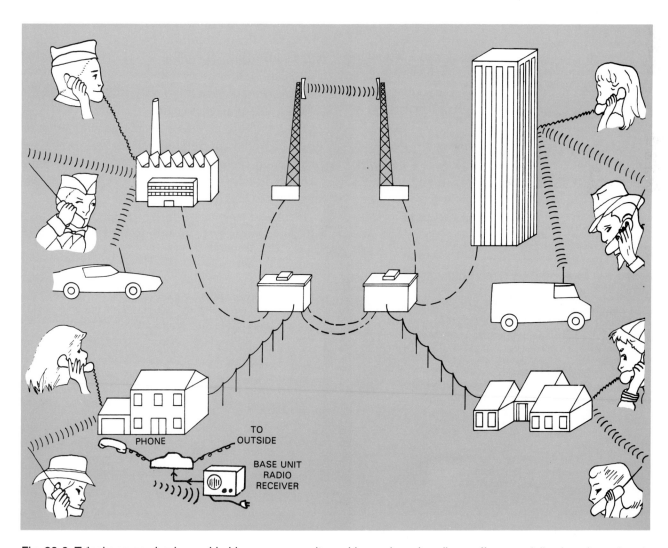

PHONE

TO OUTSIDE

BASE UNIT RADIO RECEIVER

Fig. 23-8. Telephone service is provided by a company. It provides and services lines, offers specialized equipment and services, processes calls, and transmits messages.

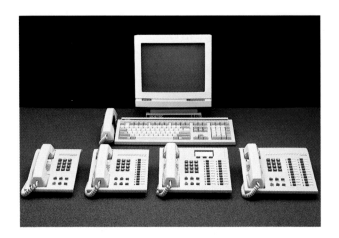

Fig. 23-9. Today, telephones do much more than help people talk with each other. They allow you to send information, transmit data to machines, and get messages when you are away.
(EXECUTONE Information Systems, Inc.)

In homes the cable enters the telephone system through a cable section called a *drop.* When many telephones are in one structure, they are wired into a terminal panel.

Telephone wires are placed overhead or underground. Overhead cables are attached to poles. These poles are often leased from the electric power company. Other poles are placed and owned by the telephone company.

Underground cable is placed in waterproof plastic conduit. Connections are made in a *pedestal.* See Fig. 23-10. Each connection is sealed into a 1 x 1/2 in. bag filled with liquid plastic. The plastic hardens in five minutes to make a waterproof covering. The bag is left in place.

In rural areas the cables and poles run along roads. In cities they are placed in an easement. An *easement* is a strip of land, usually along a property line. This strip is reserved for power, telephone, and sewer lines. Utility workers have a right to do necessary work where there is an easement.

In *fiber-optic systems,* sound waves are changed to light waves. Underground glass fiber cables are used to transmit the message. They are more efficient than copper wire cables. Many more messages can be transmitted in a much smaller cable.

A message can travel through air and space. Radio waves are used in cordless telephones. A cordless telephone can be used anywhere within a house or the surrounding average yard. It can send and receive messages from the base unit which is in your home or office. Cellular telephone systems allow you to take a portable telephone over much greater distances. Your phone messages are sent to a base station via radio waves to be relayed to their destination.

To serve a community, the modulated current is changed to short radio waves. They are called microwaves. *Relay stations* are used to send them. A relay station appears in Fig. 23-11. They are placed on towers, tall buildings, and on satellites. A relay station strengthens a radio wave and sends it to the next station or an exchange.

Two or more different telephone calls must be directed and kept separate. Calls are processed through an *exchange.* Either people or automatic switches handle the calls. They see that your message goes to the right place.

Electrical engineers design telephone systems. Most telephone systems are installed by firms that

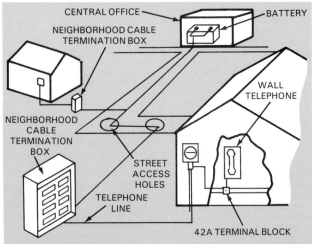

Fig. 23-10. Telephone interconnections. Left. Pedestals are used to cover spliced telephone wires and are also used for terminal covers. This pedestal cover has been removed for repairs. Right. Inside and outside wiring and hardware for telephones.

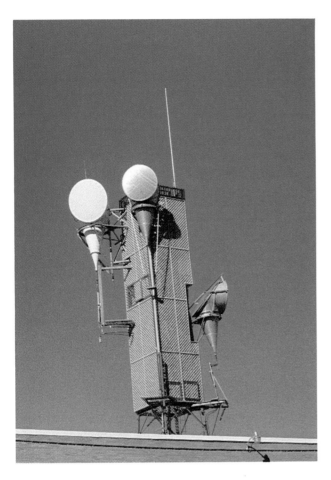

Fig. 23-11. Relay stations are placed on tall buildings.

provide telephone service or by contractors. Technicians do the installation work. The telephone users operate telephones. The owners pay for repairs on the drop and on equipment in the structure.

The telephone company provides the power to run the system. The owner buys or leases the equipment.

Television Systems

Television systems transmit sound and pictures one step at a time. The system consists of a source for the message, a video recorder/player, cables and relay equipment, and a monitor, Fig. 23-12.

Sources

A video message can be *live,* which means the message is sent and seen as it occurs. A message can be sent by a broadcasting company, or people in a firm can produce their own programs. *Video cameras* are used to convert an image into electric waves.

A message may be stored on a *videotape.* In this case, a live event is recorded. A *video recorder* imprints the electronic impulses on a tape using magnetism. The message is played back at a convenient time. Training programs commonly are stored and used this way.

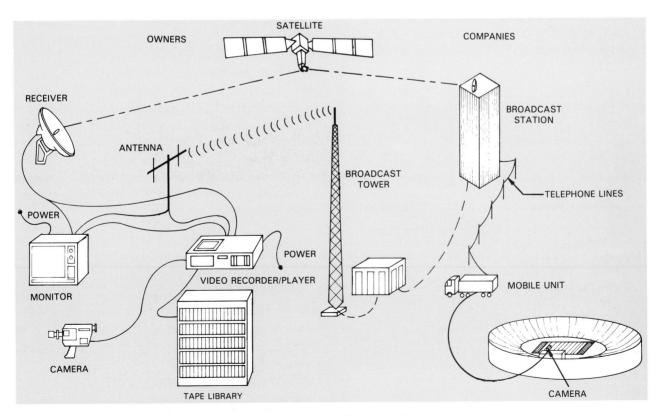

Fig. 23-12. The parts of a television system are shown. They are of about the same types for all systems. They differ in size and quality.

Video recorder/player

Video messages are sent through a video recorder/player. Pushbuttons control several functions. The recorder function makes video tapes for future use. The playback function reads recorded tapes. The message is sent to a monitor to be heard and seen.

Cables or relaying

The same kinds of cable and relay equipment used for telephones are used for television. In fact, telephone lines are used to transmit signals to the broadcast station.

Monitors

The message is heard and viewed on a *monitor.* Monitors can be small portable units. Monitors several feet high are used for large group viewing.

A television system can be portable or installed. Portable units require little construction work. Installed units are large. See Fig. 23-13. Certified technicians install the equipment.

Computer Systems

Computer systems are used to process and store information. They consist of terminals, cables, central processing units, and programs. A *terminal* is used to send and receive information. See Fig. 23-14. The information is changed into electrical impulses. The impulses are sent to the central pro-

Fig. 23-14. A terminal of a computer system is being used to design a piping layout. The layout is designed on the screen. A drawing is made by a plotter. (American Cast Iron and Pipe Co.)

cessing unit. The *central processing unit (CPU)* processes the information. It may be located inside the terminal or be miles away. For distances less than 500 ft., communication is carried on special cable called *coaxial cable.* The cable is installed between beams or studs or put in raceways. Often, two or more terminals are connected into a network. The fourth part is the program. The *program* tells the CPU what to do.

A wide variety of programs are available. Games, word processing, data management, graphics, and accounting are just a few. Browsers for the World Wide Web and electronic mail programs are available in most schools and many homes.

Two computers can communicate over a telephone line. E-mail is frequently used for sharing information with others. Confidential (secret) information is protected in these systems. A device called a modulator/demodulator (MODEM) makes it possible for telephones and computers to understand each other's messages.

Intercom System

Both the receiver and transmitter of an intercom system are owned and operated by the user. An *intercom system* is used for voice communication within a structure or complex. Large homes, apartment buildings, schools, and factories use intercoms. See Fig. 23-15.

The system uses transmitters, cables, control units, and receivers. The transmitters and receivers are turned on and off manually. See Fig. 23-16. Many kinds of systems are available.

Fig. 23-13. Television stations have equipment in racks for easy repair and revising. The equipment directs power to send the message. (Ball State Photo Service)

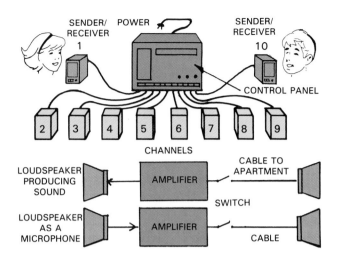

Fig. 23-15. An intercom system makes it easy to talk with people. It is used over a smaller area than telephones. Homes, schools, factories, and offices use the system.

Fig. 23-16. Some intercom systems look like telephones. (NuTone)

OTHER COMMUNICATION METHODS

One way to give information is with signs. See Fig. 23-17. Signs can include printed and molded items or electrical and electronic displays. Large signs require braced wood posts or need metal channel trusses.

SUMMARY

Efficient structures have efficient communication systems. Communication systems move or process information. Signs, bells, intercoms, television, telephones, and computers are common methods.

KEY WORDS

All the following words have been used in this chapter. Do you know their meaning?
Cable
Central processing unit (CPU)
Coaxial cable
Computer system
Drop
Easement
Exchange
Exchange system
Fiber-optic system
Intercom system
Live

A

B

Fig. 23-17. Signs are an easy way to provide information. A—Continually updated information. (Electronics Display Systems) B—Traffic control. (Richard Seymour)

Monitor
Monitoring system
Pedestal
Program
Relay station
Telephone system
Television system
Terminal
Video camera
Video recorder
Videotape

TEST YOUR KNOWLEDGE

Write your answers on a separate sheet of paper. Do not write in this book.

1. What kinds of communication systems are there? State the advantages of each.
2. When a communication system is set up to watch a place, machine, or process it is referred to as a/an _____ system.
3. Information is transmitted back and forth between people with a/an _____ system.
4. Direct conversations between people can be done with _____.
5. Outdoor telephone connections within a pedestal are often sealed against water with:
   a. Tape.
   b. Liquid plastic.
   c. Solder and caps.
   d. A housing and gasket.
   e. Electrical varnish.
6. Sound can be changed into _____ and _____ so it can be transmitted over wires and glass fibers.
7. Television systems transmit _____ and _____ signals.
8. How are monitors used in a television system?
9. Computer systems are used to _____ and _____ information.
10. What kind of system is used for voice communications within a structure or complex?

ACTIVITIES

1. How are messages sent at school? Make a sketch of the system or systems used.
2. Describe how the systems are installed. How are they operated? How would you lower their cost or plan faster installation?
3. Wire up a doorbell system for your classroom. See the plan below. A stepdown transformer is available at hardware stores.

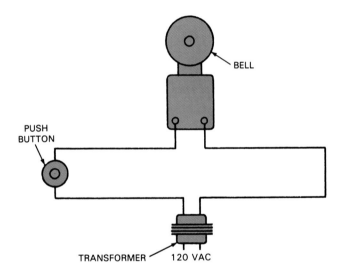

BELL

PUSH BUTTON

TRANSFORMER   120 VAC

Modern computers communicate vast quantities of information. (Texaco)

Section 7
FINISHING THE PROJECT

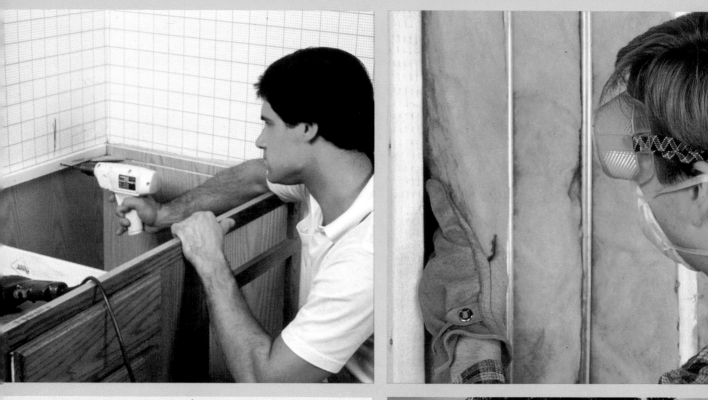

Chapter 24

INSULATING STRUCTURES

After studying this chapter, you will be able to:
☐ List three ways heat is transferred.
☐ Tell how the R-value is affected by doubling the insulation.
☐ Explain the operation of a vapor barrier and weatherstripping.
☐ List four kinds of insulation.
☐ Describe proper work clothes worn when installing insulation.
☐ Recall insulation cutting methods and two uses of glue.

We insulate structures for several reasons. Insulation helps:
• Reduce heat loss.
• Absorb sound.
• Retard burning.

This chapter describes how heat and sound are transferred. You will learn about kinds of insulation. Last, you will study about the installation of insulation.

HOW HEAT MOVES

Heat moves from warm matter to cool matter. It does it in three ways. See Fig. 24-1. They are:
• Conduction.
• Convection.
• Radiation.

When a fire burns in a stove, the outside gets hot. **Conduction** transfers the heat from the inside through the metal. Heat is conducted to your finger when you touch the stove.

Heat is conducted to the air around the stove. As it gets warmer, air begins to rise. Cooler air comes in to fill the space. The cooler air is warmed and rises, too. **Convection** currents have started. In this method the heat is transferred by moving air.

Heat is transferred through space by waves called **radiation.** That is how we get heat from the sun. Heat rays also radiate from the stove. When the waves hit our hands, they feel warm. If an object is put between the stove and your hand, the radiation is stopped. This is like being shaded from the sun. Radiation travels through empty space without heating the space. Radiation only heats the objects that it strikes.

At times we want good heat transfer. At other times we want to stop heat transfer. We want good heat transfer through the sides of the stove. We want poor heat transfer through outside walls.

Some materials insulate better than others. This ability is called **resistivity.** Resistivity is measured as an R-value. It tells how well a material resists heat movement through it. The higher the R-value, the better it insulates. See Fig. 24-2. Doubling the

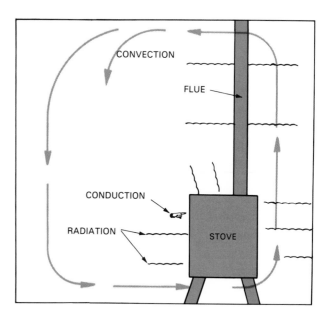

Fig. 24-1. Heat is transferred by conduction, convection, and radiation.

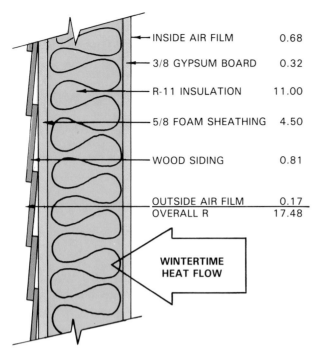

INSIDE AIR FILM	0.68
3/8 GYPSUM BOARD	0.32
R-11 INSULATION	11.00
5/8 FOAM SHEATHING	4.50
WOOD SIDING	0.81
OUTSIDE AIR FILM	0.17
OVERALL R	17.48

WINTERTIME HEAT FLOW

Fig. 24-2. The overall R-value is found by adding up the R-value of each layer in the wall. Heat flow depends on R-values, not the thickness.

thickness of insulation roughly doubles the R-value. Wood is a better insulator than concrete. Foam plastic is better than wood.

HOW SOUND TRAVELS

Sound is energy waves. They travel in air and in all directions from the source. When the energy waves hit walls, floors, or ceilings, some of the energy is reflected. The rest is absorbed. Reflected sound can be heard as an echo. Absorbed sound goes through a barrier. It is heard on the other side. See Fig. 24-3.

Acoustical engineers are concerned with reverberation characteristics (echoes) and sound transmission. *Reverberation characteristics* relate to how long it takes for sound to be muffled in a room. Carpets, furniture, and people in a room reduce reverberation. However, muffling sound in a room does not reduce sound transmission.

Sound transmission occurs when sound moves through a barrier. Sound transmission is reduced by increasing the barrier's mass. Masonry walls and more than one layer of gypsum board increase mass. *Staggered stud construction* separates the moving surfaces. *Sound-deadening materials* cushion them. Fig. 24-4 shows four ways to reduce sound transmission.

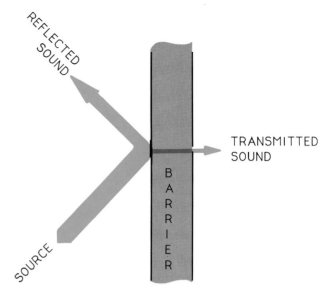

Fig. 24-3. Some sound is reflected back into the room, and some is absorbed by the barrier.

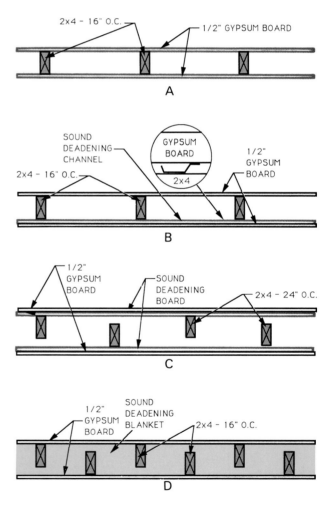

Fig. 24-4. Methods of decreasing sound transmission. A–Increasing the mass. B–Using a sound-deadening channel. C–Using staggered stud construction and sound-deadening board. D–Using staggered stud construction and sound-deadening blanket. Any combination of these techniques also reduces sound transmission.

THERMAL INSULATION

Insulation is used to reduce heat transfer. It holds heat in during the winter. In the summer, insulation keeps heat out.

What Insulation Is

Fluffy materials that trap air are used to reduce conduction. See Fig. 24-5. Solid glass is a poor insulator. When glass is spun into fine fibers it becomes *fiberglass.* Fiberglass is mostly air and is a good insulator. Wood fiber, cork, and foam plastic also have a lot of air in them. They are all good insulators against conducted heat.

A

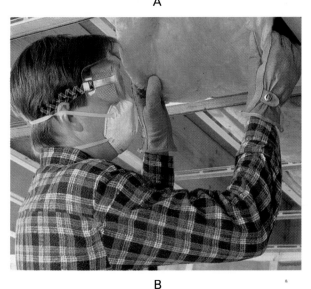

B

Fig. 24-5. Glass is spun into small fibers. The fluffy material will trap and hold air. A–Heat will not pass through it easily. B–Acoustical batts and sound-deadening channels reduce sound transfer. (Owens-Corning Fiberglas Corp.)

Convected heat is lost through air currents. Either unwanted air gets into, or heated air gets out of, the structure. The first case is called *infiltration.* Infiltration is slowed by sealing cracks. The cracks may be found around doors, windows, and the foundation. *Weatherstripping* is used around doors and windows to stop air flow. *Caulking* is used to close other cracks in the structure. Storm windows and magnetic seals on doors seal buildings even tighter.

There are several ways to reduce infiltration. The entire outside of the structure may be wrapped, Fig. 24-6. An airtight plastic fiber film is used. This film lets moisture through while retaining heat. The inside gets a similar film covering. It does not let moisture into the insulation. In new construction, walls, ceilings, and floors are covered before gypsum board or plaster is installed.

Through sealing and wrapping, structures can become too airtight. Fresh air cannot get into very tight structures. The air inside can become polluted. If tight seals everywhere are required, then *air exchangers* are used to bring in fresh air and to save energy. The air being removed is used to adjust the temperature of the air coming in.

Types of Insulation

Insulating materials are of four main types. They are:
- Rigid boards.
- Batts and blankets.
- Loose fill.
- Reflective.

Fig. 24-6. Air infiltration is reduced by wrapping the building with a plastic fiber sheet. Air cannot get through. Moisture can. (TYVEK by DuPont)

All except reflective insulation work in the same way. They trap air in tiny pockets inside the material. The air slows heat transfer in both directions.

Rigid boards are made of plant materials or foam plastic. Foam plastic has the highest R-value. *Batts and blankets* fill the cavities between framing members. See Fig. 24-7. They prevent convection currents inside walls. They also reduce conducted heat. *Loose fill* is poured or blown in place. See Fig. 24-8. It comes in bags. Special machines are used to place it.

Reflective insulation works on a different principle than the others. It has foil surfaces that stop heat waves. The shiny surface works like a mirror. The heat wave hits it and bounces back. Note that if the surface was black, the heat waves would be absorbed instead of reflected.

Shiny foil is not always used to insulate. It may serve to keep moisture from getting into the insulation. See Fig. 24-9. Foil is put on the facing side (inside) of insulation placed between framing members. The backs of gypsum board are sometimes covered with foil.

Vapor Barriers

Air has water vapor in it. The amount of water vapor in air is called the relative humidity. Water vapor moves in the same direction as heat. Refer to Fig. 24-10. It travels from the warm side of a wall to the cool side. In the winter, moisture moves from the inside to the outside. As it passes through the wall it gets cooler. Fig. 24-11 shows what happens. When the air cools, the water vapor condenses (turns to liquid water). The water makes the insulation wet. The air spaces begin to fill with water. The

Fig. 24-8. Old newspapers, melted rock, and glass can be made into a fluffy material. The material is used as loose insulation in walls and attics. (CertainTeed Corp.)

Fig. 24-9. The shiny foil serves as a vapor barrier. (CertainTeed Corp.)

A

B

Fig. 24-7. How batts and blankets are held in place. A–Unfaced fiberglass batts are cut slightly larger so friction holds them in place. B–Blankets have a kraft-paper facing. Staples hold them in the walls. (CertainTeed Corp.)

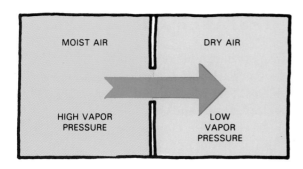

Fig. 24-10. Vapor pressure causes water vapor to move to drier areas. (Manville Building Materials)

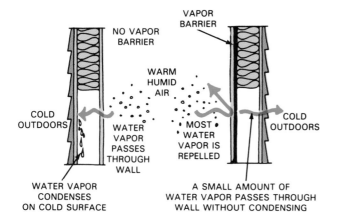

Fig. 24-11. A vapor barrier stops most of the moisture before it gets into the wall.
(Manville Building Materials)

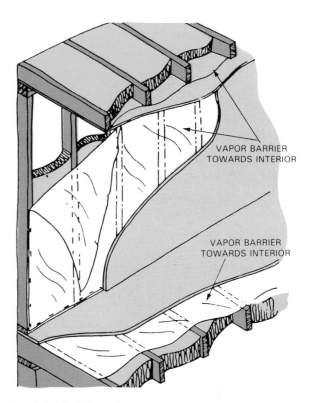

Fig. 24-12. Vapor barriers are placed facing the inside of the structure. (Manville Building Materials)

Fig. 24-13. This paint prevents water from getting into the basement. Water vapor cannot get out either. (DRYLOK by United Gilsonite Laboratories)

wet insulation is not a good insulator and lets too much heat out. The water can also cause the frame members to rot and cause paint to peel.

A *vapor barrier* stops the moist air before it changes to water. A plastic sheet placed under the wall covering serves as a vapor barrier, Fig. 24-12. Wall coverings or the insulation may be backed with foil. The foil will hold back the water vapor. Vapor barriers should always be placed on the warm side of the wall.

There are vapor barriers that are applied as paint. Most often, this is an aluminum paint. It is put on with brushes, rollers, or sprayers. The paint is used when an old house is being insulated. After insulation is blown into an existing wall, the inside of the wall is painted. In this way, a plastic sheet barrier is not needed.

There are also paints that keep out standing water, Fig. 24-13. Some are used as vapor barriers. Because these paints can keep out standing water, they can also block the smaller amounts of water due to vapor. They are used on concrete block basement walls. Again, a plastic sheet barrier is not needed.

What to Insulate

All outside walls, ceilings, and floors should be insulated. See Fig. 24-14. Sometimes pipes and junction boxes are on outside walls. Insulation should be placed between the pipes or boxes and the sheathing, as in Fig. 24-15. Pipes carrying hot or chilled liquids are insulated, Fig. 24-16.

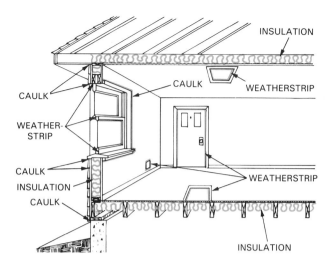

Fig. 24-14. All outside walls, floors, and ceilings should be insulated.

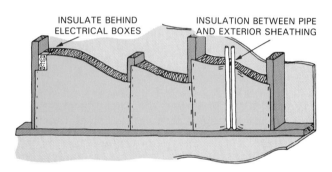

Fig. 24-15. Insulation behind junction boxes cuts down drafts. Pipes will not freeze when insulation is between the pipes and the sheathing. (Manville Building Materials)

Fig. 24-16. Specially formed fiberglass tubes are used to cover straight and curved pipes. (CertainTeed Corp.)

INSTALLING INSULATION

Insulation can make you sick or can cause your body to itch. Proper work clothes can prevent most problems. Fig. 24-17 includes suggestions. Wear a long sleeved shirt and gloves. They keep the insulation off your skin. Goggles and a respirator keep the insulation out of your nose and mouth. Loose-fitting clothes make it easier to work. Tight clothes may rub the insulation into your skin.

Shower with soap and water after working with insulation. Wash work clothes separately from other clothes. Then, rinse the washer before using it again.

Placing Loose Insulation

Loose insulation is put in place by pouring and blowing. The material comes in bags. See Fig. 24-18. It can be poured into spaces between walls and poured into attics. Some materials, like loose plastic beads and vermiculite insulation, have concrete mixed with them. They are poured on roof decks to insulate ceilings.

Special blowers are used to place loose insulation in walls and attics. See Fig. 24-19. The blower fluffs the insulation, then blows it into place.

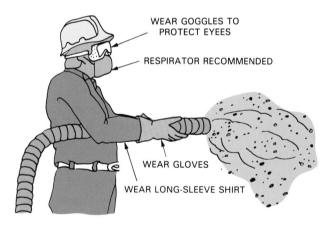

Fig. 24-17. Workers must be careful when they work with insulation.

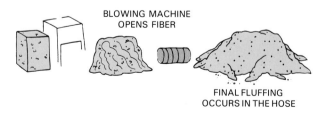

Fig. 24-18. Loose insulation is compressed into bags. A machine fluffs it up. (Manville Building Materials)

Fig. 24-19. Loose insulation is often used in attics. A blower on a truck moves the insulation through a hose to the attic. (CertainTeed Corp.)

Placing Blankets and Batts

Blankets and batts are designed to fill the voids between framing members. You can buy standard widths for wood frame structures. They fit between frame members set (spaced) on 16 and 24 in. centers. Batts are 48 to 96 in. long. They are held in place by friction between the frame and the batts.

Blankets have a paper backing. Rolls are up to 24 ft. long. The paper edge is stapled to the framing.

Sometimes special methods are needed to hold insulation placed in ceilings or under floors. An example is a crawl space, where the vapor barrier must be up toward the heated space. The paper edges are not available for easy stapling. Wire in many forms can be used to hold insulation up. See Fig. 24-20.

A knife works well for cutting insulation batts and blankets, Fig. 24-21. A knife with a sawlike edge is best. A hammer, stapler, measuring tape, straightedge, and ladder make the job easy.

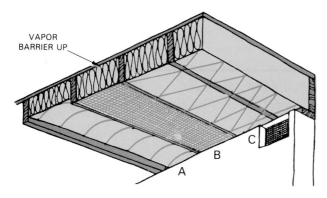

Fig. 24-20. Insulation in floors has to be held in place. Special fasteners (A), wire mesh (B), and wire lacing (C) are often used.

Fig. 24-21. Measure carefully when cutting insulation blankets so they will fit tightly. (Owens-Corning Fiberglas Corp.)

Placing Rigid Insulation

Rigid insulation is nailed or glued in place. It is nailed to the outside of the framing as sheathing. See Fig. 24-22. An air space should be left between foil-faced boards and the siding. The air space will reduce conduction.

Placing Foam Boards

Foam plastic boards resist water. They are used to insulate foundations and concrete floors, Fig. 24-23. Glue or friction holds them in place.

Fig. 24-22. Foam plastic sheathing is a good insulator. This foam is foil-backed. (SENCO Products, Inc.)

Fig. 24-23. Foam plastic board is used to insulate the foundations of buildings.

SUMMARY

Heat moves by conduction, convection, and radiation. Insulation is used to reduce heat loss, absorb sound, and retard burning. The types of insulation include rigid boards, loose fill, batts and blankets, and reflective. A vapor barrier keeps moisture from condensing on insulation and from ruining wood and paint.

Loose insulation is placed by pouring, blowing, and spraying. Blanket and batt insulation is held in place with friction, staples, and wire. Nails or glue hold rigid board insulation in place.

KEY WORDS

All of the following words have been used in this chapter. Do you know their meaning?

Acoustical engineer
Air exchanger
Batts and blankets
Caulking
Conduction
Convection
Fiberglass
Infiltration
Loose fill
Radiation
Reflective insulation
Resistivity
Reverberation characteristic
Rigid board
Sound
Sound transmission
Sound-deadening materials
Staggered stud construction
Vapor barrier
Weatherstripping

TEST YOUR KNOWLEDGE

Write your answers on a separate sheet of paper. Do not write in this book.

1. Which wall will let more heat pass through it, one with a resistivity of R-19 or R-38?
2. How is heat being transferred when warmed air moves from a warm place to a cool place?
3. How do fluffy insulation materials reduce conduction?
4. How can you reduce infiltration?
5. How can you stop radiant heat from passing through a wall?
6. Which one of the following is not a type of insulation?
   a. Loose fill.
   b. Thermoplastic.
   c. Batt/blanket.
   d. Reflective.
   e. Rigid board.
7. Explain why you must have a vapor barrier in the outside wall of a heated structure.
8. How do you place loose insulation?
9. Where are blanket and batt insulation placed?
10. How is rigid insulation held in place?

ACTIVITIES

1. How can you insulate yourself from the heat in the following cases:
   a. Standing in the sun.
   b. Standing barefoot on asphalt in the sun.
   c. Being in the stream of a hair drier.
   Try out some of your solutions.
2. Calculate the total R-value of the following layered structure: 8 in. of fiberglass batting, 1/2 in. of gypsum board, and 1/8 in. of hardwood paneling. Common R-values are:
   • Fiberglass 1 in. 3.13.
   • Gypsum 1 in. 1.00.
   • Hardwood 1 in. 0.91.
3. Test three concrete blocks for water leakage after using paints to seal two of the samples. First fill the core with mortar 1 in. thick to make a container out of the core hole. Then use two types of paint on the outside of two blocks. Let them dry. Fill all three blocks with water to the top. Count how long it takes for water to first bleed out and also to empty out. Make a chart of your results.

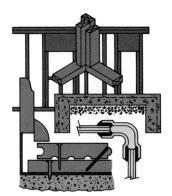

Chapter 25
ENCLOSING THE INTERIOR

After studying this chapter, you will be able to:
- ☐ Identify bearing walls and partition walls.
- ☐ List the three types of plaster coats.
- ☐ Describe the cross section of a joint in gypsum board.
- ☐ Explain what is done to prepare masonry walls for coverings.
- ☐ Describe ceiling types.
- ☐ State the purpose of subflooring and underlayment in structures.
- ☐ Explain what is done to make concrete floors wear-resistant.

Ceiling, wall, and floor coverings are needed for three reasons:
- Appearance.
- Sound insulation.
- Heat insulation.

The coverings hide the framing, pipes, ducts, wires, and insulation. Interiors are covered with gypsum board, plaster, tile, and paneling.

Homes, offices, and schools need to be completely finished inside. See Fig. 25-1. Factories may not be finished inside at all. Some are not even enclosed, Fig. 25-2.

Enclosing the inside involves doing rough work and finish work. Rough work is covered by the finish work. Rough work is described in this chapter. Finish work is discussed in the following chapter.

WORKERS WHO ENCLOSE INTERIORS

The architect works very closely with the owner. They decide on materials, colors, and textures. General contractors see that the work is done. Much of the work is done by subcontractors. Carpenters and masons often work for the contractor.

Fig. 25-1. The ceiling covers pipes, ducts, wires, and framing. The redwood walls cover pipes and wiring. The floor is soft, reduces noise, and improves the appearance. (Calfornia Redwood Assoc.)

See Fig. 25-3. Iron workers, glaziers, and plasterers are hired as subcontractors. Floor covering work is also subcontracted. See Fig. 25-4. Subcontractors need special skills and tools to do their best.

ENCLOSING WALLS

Outside walls enclose the space within a structure. Inside walls divide the space. *Bearing walls* are inside walls that help support the structure. *Partition walls* only divide space. The inside of exterior walls and both sides of partition walls are covered. Partition walls are usually installed first. Then the ceilings are covered.

Plaster Walls

In older structures, lath and plaster were used to cover wall frames. *Plaster* is a mixture of sand,

A

B

Fig. 25-2. A—The framing, wiring, and pipes are left exposed in factories. (TransLogic Corp.) B—In refineries, little equipment is enclosed. (EXXON Corp.)

Fig. 25-3. There is a lot of work on most projects for carpenters. The contractor hires them.

Fig. 25-4. Special skills and materials are needed by people who lay flooring. They usually work for subcontractors. (Armstrong World Industries, Inc.)

lime, and water. *Lath* are narrow strips of rough wood. The lath were spaced so that soft plaster could squeeze between them. When the plaster hardened it held tightly.

Wood lath has been replaced with wallboard or metal lath. *Wallboard,* also called gypsum lath, replaces the first coat of plaster.

Sections of metal lath are fastened to the framing. *Metal lath* has many holes in it. Part of the first coat of plaster is pushed through the holes. This coat is called the *scratch coat.* A second coat is called the *brown coat.* The brown coat builds up the thickness and levels out the first coat. The *finish coat,* Fig. 25-5, is applied last. The finish coat leaves a smooth, even surface. Plastering is done by highly skilled workers.

Gypsum Board Walls

Today, **gypsum board** (drywall) is used more than plastering. It comes in 4 ft. wide sheets. They are strong and smooth and go up fast. See Fig. 25-6. Gypsum board is fastened with nails, screws, or mastic. *Mastic* is a thick glue for mounting the panels to studs.

Joints and nail dents are covered with *joint compound,* Fig. 25-7. A *paper tape* is added to the joints. A second layer of joint compound is added and smoothed. It is left to dry. The rough spots are sanded lightly, and a third coat is applied. After it dries, the joints and nail dents are sanded or sponged. A damp sponge smooths the joint compound. It is now ready to paint.

Concrete board resists water. This material is a sheet of concrete 1/2 in. thick. It is reinforced with

Fig. 25-5. These workers are putting on the finish coat that is smooth and even. Note that one worker is using stilts. (National Plastering Industry's Joint Apprenticeship Trust Fund)

Fig. 25-6. Gypsum board goes up fast and is low cost. Edges are tapered to allow for buildup of joint compound. (Owens-Corning Fiberglas Corp.)

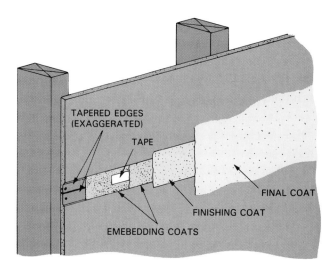

Fig. 25-7. Taping joints makes them stronger. It keeps them from cracking. Allow the joint compound to dry, and sand it after each step.

a fiberglass mesh on both sides to make it strong. Concrete board or water-resistant gypsum board is used in damp places such as shower stalls and locker rooms. Ceramic tile or plastic coverings must be used.

Masonry Walls

On masonry walls, furring strips are needed before applying drywall or other wall material. See Fig. 25-8. *Furring strips* are narrow pieces of wood. They are fastened with masonry nails which are made of hardened steel. The nails can be driven into concrete. Insulation can be placed between the furring strips.

Fig. 25-8. Wood furring strips are fastened to concrete block walls. They provide space for insulation and a nailing surface for paneling or gypsum board. (CertainTeed Corp.)

If a masonry wall is going to be covered with paneling or some other sheet, furring strips and two vapor barriers are often used. Place furring strips as shown in Fig. 25-9. Final vertical alignment can wait until the paneling work begins.

ENCLOSING CEILINGS

Ceilings are the overhead surfaces in rooms. Some ceilings are the underside of the roof. Other ceilings have their own framing well below the roof structure.

Plaster and Gypsum Board Ceilings

Plaster and gypsum board ceilings are installed before the walls are covered. The same skills, materials, and methods are used for both.

Ceiling Tiles

Ceiling tiles are small pieces of ceiling material. They are usually 12 in. x 12 or 24 in. Their thickness varies. Ceiling tile can be bonded to a flat surface with mastic. Mastic is used when nails or staples cannot be used.

Furring strips for ceiling tiles are narrow 1 in. thick boards. The strips are nailed to joists overhead. See Fig. 25-10. The furring strips are spaced

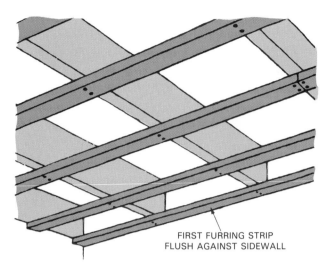

FIRST FURRING STRIP
FLUSH AGAINST SIDEWALL

Fig. 25-10. Furring strips are nailed to ceiling joists. The ceiling tile is stapled to them. (Armstrong World Industries, Inc.)

to support the tiles. Staples attach the tile to the strips.

Suspended Ceilings

Suspended ceilings are used to cover parts of the mechanical system like pipes, electrical wiring, or heating ducts. They may be installed to lower the ceiling height. Wire hangers are attached to the framing above. See Fig. 25-11. The grid (frame) system is fastened to the wires. Ceiling panels and

A

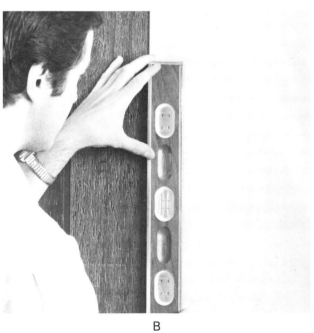

B

Fig. 25-9. On masonry walls, a vapor barrier can be put between the concrete blocks and the furring strips. A second (main) barrier must be on the inward side of insulation. A–Careful carpenters plumb the first furring strip and then use a spacer strip to align other furring strips. B–The first panel should be exactly plumb. (Masonite Corp.)

Fig. 25-11. A laser beam is used to level this grid system. Ceiling panels and lights will be held up by the grid work. (Spectra-Physics)

sometimes lights are set into the grid system. Commercial systems may include lights and heat ducts. Fig. 25-12 shows one of many.

INSTALLING FLOORS

The floor is the last surface of a room to be finished. It receives two covers. One is called the subfloor. The other is the finish floor.

Subfloors

The subfloor frame structure is installed early. A *subfloor* provides weight-bearing strength. It is used to store materials and as a work space. In large structures and basements, the subfloor is made of concrete. In wood frame structures, it is plywood, 1

x 6 lumber, or flakeboard. Fig. 25-13 shows a flakeboard subfloor.

A plywood subfloor is often covered with underlayment. See Fig. 25-14. *Underlayment* provides a smooth surface for floor coverings. Smoothed plywood, particle board, and hardboard are common materials.

Concrete Finish Floors

Floor surfaces need to meet many demands. Factory floors must resist heavy traffic, oil, and other heavy use. Commercial floors require other features. They must be good looking, easy to clean, and wear well.

Fig. 25-13. Flakeboard is strong and smooth. Some products have tongues and grooves on their edges. They hold even. (SENCO Products Inc.)

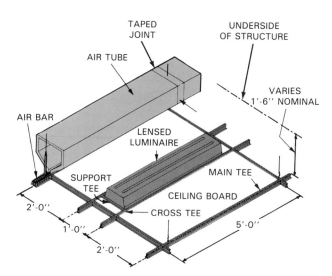

Fig. 25-12. Engineered ceiling systems have lights and air ducts in them.
(Owens-Corning Fiberglas Corp.)

Fig. 25-14. Underlayment is smoother than unsanded plywood. A thin flooring material will be laid over this underlayment. (MacMillan Bloedel Ltd.)

Concrete floors are used in commercial and industrial structures. A layer of special concrete is placed over a freshly poured concrete slab. It is troweled smooth. Hard aggregate and short steel wires or iron beads are added to the concrete mix for some factory floors. These floor surfaces wear longer than plain concrete.

SUMMARY

The walls and ceilings are enclosed first. The floors are done last. Where people will see the work, skilled workers enclose the inside of structures. Plaster, gypsum board, and panels are used. These wall systems may use lath or furring strips.

Ceilings are put up before walls are covered. Some ceilings are not enclosed. Others are enclosed with plaster and gypsum board like the walls. Tile, grid work, and engineered (grid work) systems are used to enclose ceilings.

Floors consist of a subfloor and finished floor. Underlayment is often put over the subfloor. Floors with reinforced concrete surfaces are wear-resistant.

KEY WORDS

All of the following words have been used in this chapter. Do you know their meaning?

Bearing wall
Brown coat
Ceiling
Ceiling tile
Concrete board
Finish coat
Furring strip
Gypsum board
Joint compound
Lath
Mastic
Metal lath
Paper tape
Partition wall
Plaster
Scratch coat
Subfloor
Suspended ceiling
Underlayment
Wallboard

TEST YOUR KNOWLEDGE

Write your answers on a separate sheet of paper. Do not write in this book.

1. Inside walls that help support the structure are called _____ _____.
2. When taping gypsum wallboard joints, no fewer than _____ coats of joint compound are applied.
   a. two
   b. three
   c. four
   d. five
3. The three coatings of plaster are called _____, _____, and _____.
4. In many cases, plaster has been replaced with _____ board.
5. Gypsum joints are sealed with _____, _____, and _____.
6. A moisture resistant board called _____ _____ is used to enclose walls of showers.
7. Often, masonry walls have wood _____ _____ fastened to them with masonry nails.
8. Wiring, pipes, and ducts can be covered with _____ ceilings.
9. A wood subfloor is made smooth for thin floor coverings with _____.
10. Describe how a furniture factory floor and an auto transmission factory floor might differ.

ACTIVITIES

1. Check how the floors in your home were made. Select a floor covering for a room of your choice. It can be a clubhouse, your bedroom, or another room in your house. What kind of subflooring do you need for the covering? How is it installed?
2. How would you enclose the walls in your garage or clubhouse? What tools would you need to buy to do it yourself?
3. List the steps to cover a wall in your garage with gypsum board. Describe ways to put up a ceiling in a room.

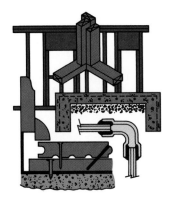

Chapter 26

FINISHING THE PROJECT

After studying this chapter, you will be able to:
☐ Describe several finishing trades and designing professions.
☐ Describe paint types and painting methods.
☐ List three materials for covering walls.
☐ Tell how to lay floor tiles, hardwood flooring, and resilient flooring.
☐ Choose tools used to cut moldings.
☐ State the uses of caulking materials.
☐ Compare rough-in work with finishing work on utilities.
☐ Recognize types of temporary walls.
☐ Explain how to lift furniture properly.

Finish work in a house, on a highway, or at a plant is done last. Both the inside and the outside of a structure are finished. Finish work includes four jobs: trimming, painting, decorating, and installing. Painting protects the structure from wear and weather. A painted surface is easy to maintain. A surface may be covered with something other than paint. This is called decorating. Trimming covers cracks and seals the structure from weather. Installing involves placing, connecting, and testing equipment and fixtures. An improvement in appearance is a goal of all four finishing jobs.

WHY FINISH WORK IS DONE

Finish work is done for three reasons:
• Appearance.
• Protection.
• Usefulness.
Appearance is of more value in an office building than in a factory. Finished projects are protected against weather and use. They are not drafty nor do they leak water. Finished structures last longer. Finish work is done to make the project more use-

ful. Installed equipment makes the building easy to use. For examples of finish work, refer to Fig. 26-1. With installed items and protected surfaces, the climate is pleasant and the surfaces are easy to maintain.

WHO DOES FINISH WORK

Architects, engineers, and interior designers select the materials, fixtures, and equipment. Architects design the outside. Interior designers select colors, materials, and finishes for the inside.

Interior designers have a background in art rather than construction. They use their skills to plan space. They combine colors, texture, and form within the space. The insides of structures are made more useful and attractive, Fig. 26-2.

Some interior decorators may work for themselves. Others may work for architects or retail stores. Homes are often decorated by the owner. The professionals all work closely with the owner. See Fig. 26-3.

Subcontractors often do the finish work. Painting, decorating, and floor covering subcontractors follow the working drawings. Equipment and furnishings may be installed by the suppliers. Trimming is often done by the general contractor.

FINISHING EXAMPLES AND METHODS

Finishing includes four jobs:
• Painting.
• Decorating.
• Trimming.
• Installing.
Materials are covered and protected by painting and decorating. Trimming closes up cracks found in corners and around openings and helps dress up

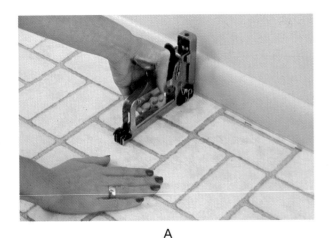

A

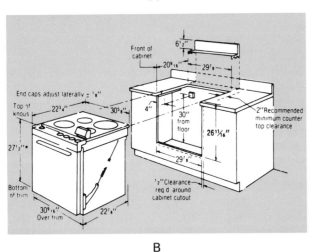

B

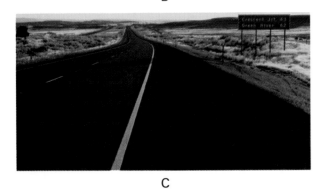

C

Fig. 26-1. Finishing work takes place on all types of construction projects. A–Floor coverings beautify floors, are easy to clean, and protect the structure beneath. (Armstrong World Industries, Inc.) B–Installing may include building or placing cabinets for appliances. (Frigidaire Corp.) C–Finishing shoulders, erecting signage and reflector posts, and painting stripes are all finishing work done on highways. (Utah Department of Transportation)

windows and doors. Caulking is a low cost way to seal cracks between surfaces. Floor coverings can include wood, carpet, plastic, and tile. Installing is work on equipment and furnishings and may occur throughout construction. See Fig. 26-4.

Fig. 26-2. An interior decorator chose colors, textures, and forms. Painters, carpenters, carpet installers, and electricians finished the work. (Pella/Rolscreen Co.)

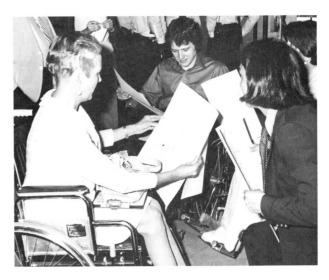

Fig. 26-3. The comfort of many people must be considered when designing the inside of a building. What conveniences might a person in a wheelchair require in a home? (President's Committee on Employment of the Handicapped)

Painting

Painting helps protect and enhance a structure. Inside, painting is often the first finish work to be done. It is faster to paint before the trim is up and outlet covers are put on.

Why paint?

We paint for four reasons:
- To protect.
- To add color.
- To make cleaning easier.
- To make space more pleasant.

Fig. 26-4. How many of the four finishing processes were used to finish this office area? Can you list the jobs and put them in the order that they may have been done? (California Redwood Assoc.)

Painted wood will stay clean longer. Undercoated and painted steel does not rust like bare steel. Paint adds color, and it can create a mood. Light colors (those mixed with white) make a room seem larger. Darker colors (less or no white) make a room look smaller. Painted surfaces are easy to clean. A damp, soapy cloth can usually make the color bright again.

Paint ingredients

All paint has three ingredients. They are the vehicle, pigment (color), and thinner. Special paints have more ingredients.

The *vehicle* is the coating material. Oil and latex are the most common. Oil-based paints need special cleaners and thinners. Water is used to clean up and thin latex paint. Usually, oil paint lasts longer, but it costs more.

The color is a *pigment* (solid material). You can buy paint in standard colors. Many hardware and paint stores can mix hundreds of colors.

Thinner controls the thickness of the vehicle. When the thinner evaporates (leaves), paint is dry.

Finish

Paint has a *finish sheen,* or luster. It ranges from a flat (dull) finish to a gloss (shiny) finish. A semigloss is in between. Flat paint appears soft. It is used to reduce reflections. However, it is hard to clean. Gloss paint is easy to clean and wears better, but it has a glare. Semigloss paint has some qualities from both types of paint.

Lacquer and varnish

Lacquer and *varnish* are transparent protective finishes for woods. *Stains* add color, help make wood color more consistent, and let most of the grain show through. Varnish or lacquer coats are applied over the stain to preserve the stain and protect the wood. They make the surface more durable and easier to clean. Lacquer and varnish allow wood grain pores (natural satin appearance) to show through.

Spreading

Paint is spread with brushes, rollers, and sprayers. Brushes are used on small areas. Rollers are used on larger ones. Sprayers are much faster than brushes or rollers. See Fig. 26-5.

Preparation is needed before painters can begin. When they begin work, they prepare the surface. Dirt is removed. Cracks and holes are filled and rough spots are smoothed. A primer seals gypsum board. See Fig. 26-6. Preparation is the only way to get speed and quality in paint work.

Masking tape, paper, and drop cloths are used to keep paint off floors, bushes, and windows. Again, preparation is very important to save work.

Painters work from ladders and scaffolding. Scaffolding may be freestanding, attached to ladders, or attached from above. Electric motors raise and lower those hanging from ropes or cables.

Street painting

Street painting involves large areas and thick, tough paints. Typical projects are shown in Figs. 26-7 and 26-8. A stencil or template is sometimes used for making patterns. A sprayer is the best way to save time on great lengths of highways.

Decorating

Decorating walls can be done with wallcovering, paneling, boards, and tile. Wallcovering includes wallpaper, vinyl covering, foils, and fabrics.

Wallpaper

Wallpaper comes in rolls. Lengths of material are cut to fit the surface. A paste is used to hold it on the wall. The paste may already be on the paper. If not, most paste can be brushed or rolled onto the paper. After putting paste onto the paper, the paper is folded paste side to paste side, Fig. 26-9. This makes it easier to handle. Most paste will not stick to itself very well. Thus, folding the paper will not damage it or cause problems.

A

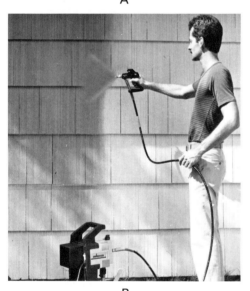

B

C

Fig. 26-5. Painting tools. A—Applying high gloss oil paint with a brush. (Mike Brian, Indiana & Michigan Electric Co.) B—Power paint sprayer. C—Power roller. (Wagner Spray Tech Corp.)

Fig. 26-6. The surface is sealed with a primer. This saves paint during the final coat.
(Domtar Industries, Inc.—Upson Products)

Fig. 26-7. Heavy duty paint is used to mark highways. A template is used to paint shapes.
(Muncie Newspapers, Inc.)

Fig. 26-8. This machine makes sharp edges on the stripes. The worker needs no tape or template. (Kelley-Cresswell Co., Inc.)

Fig. 26-9. Fold the paper to touch paste side to paste side. Some papers must set at this stage for a few minutes before hanging. (Glidden)

When the wallcoverings are prepasted, the roll must be put into a water tray to activate the paste. Then, the covering is simply unrolled upwards to mount at the ceiling.

The paper is placed on the wall. Care is taken to keep it straight up and down. Patterns are carefully matched at the edge.

Paneled walls

Paneled walls cost more than gypsum board walls. Most paneling is made of thin sheets of plywood or hardboard. Hardboard is made of processed wood fibers put into a mat and compressed under heat and pressure. Textured rollers can put many patterns onto the panel. A woodgrain pattern

Fig. 26-10. The walls in this room are covered in hardboard paneling. The pattern is woodgrain. (Masonite Corp.)

and texture can be printed on the hardboard, Fig. 26-10. A thin sheet of vinyl is rolled on the paneling as a finish.

Plywood, Fig. 26-11, is made up of thin layers of wood. The grain in each layer runs at right angles to the next layer. The layers are glued together to form a rigid panel. The panel is sanded and finished. Fig. 26-12 shows one type of finish. Paneling is easy to

Fig. 26-11. Plywood paneling is light and strong. It does not require painting.
(Owens-Corning Fiberglas Corp.)

Fig. 26-12. These plywood panels are called architectural plywood. The architect selected the veneer on the plywood. The panels are custom-made just for this room. (Mutual Federal Savings Bank)

install with either nails or mastic. It is fastened directly to framing, gypsum board, or furring strips.

Solid wood paneling is the most costly. The fewer defects the wood has the more it costs. Nails and mastic are used to hold boards in place. See Fig. 26-13. A tongue-and-groove joint is used to keep the boards aligned.

Fig. 26-13. Solid California redwood is a long-lasting wall covering. It adds warmth to a room. (California Redwood Assoc.)

Tiles

Ceramic and plastic tiles are used in wet areas. See Fig. 26-14. Plastic tile costs less, but it does not last as long. Tile does not absorb water and will not rot. It is easy to clean. Tile can be set on moisture resistant (MR) gypsum board. Concrete board is better if there is a lot of water.

Grout is a material, like plaster, used to fill cracks between the tiles. It can be made in colors to match or contrast with the tiles.

Fig. 26-14. Ceramic tiles are used in showers and bathrooms. Tiles resist water, wear well, and are easy to clean. (Pella/Rolscreen Co.)

Wall treatment systems

Wall treatment systems are those that include special ways to attach panels to a wall. A range of features and colors are often available. Sizes are 32 in. by 8 ft. or 48 in. by 8 ft.

The system in Fig. 26-15 uses metal strips to hold it to the wall. The first strip is attached to the wall. The strips fit into the edges of the panel. The next strip is slipped into place and is fastened to the wall. The work goes on until the wall is covered.

Finishing Floors

Floors are finished with rigid masonry materials like ceramic tile, stone, or terrazzo or with resilient coverings like wood or carpet. See Fig. 26-16.

Ceramic tile, stone, and *terrazzo.* Ceramic tile floors wear well and look good. Tiles and stone are set in a layer of mortar. Grout is worked into the cracks between the pieces. Stone floors are laid in a similar method.

Terrazzo floors use a special concrete. Two layers of special concrete are placed on top of a concrete slab. See Fig. 26-17. The first layer is a sand

Fig. 26-15. This wall treatment is easy to install. The board is made of wood fibers. The surface controls sound in a room and between rooms. (Owens-Corning Fiberglas Corp.)

mortar mix. Metal strips are set on edge in the mortar. After the mortar hardens, a terrazzo mix is poured. *Terrazzo* consists of white portland cement, coloring, sand, and marble chips. This mixture is placed, smoothed, packed with a roller, and left to harden. It is then ground, polished, and sealed.

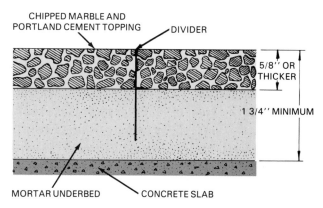

Fig. 26-17. Terrazzo is used on floors that must look good, last a long time, and clean easily.

Resilient flooring. ***Resilient flooring*** is made of plastic and fibers. It comes in rolls or squares. Refer to Fig. 26-18. The material gives a little when you walk on it. It comes in many colors and patterns, is

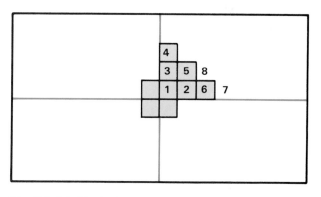

Fig. 26-18. Typical sequence for laying resilient floor tile. This is the pyramid pattern method.

A

B

Fig. 26-16. Flooring materials. A–Carpeting. (Elliot Corp.) B–Ceramic tile for floors is made of clay. It is covered with powdered glass. When it is fired, the glass melts and bonds to the clay. (California Redwood Assoc.)

easy to clean, and wears well, Fig. 26-19. It is held in place with an adhesive. Resilient flooring can be installed over a smooth concrete or wood subfloor, Fig 26-20.

Wood flooring. Wood flooring comes in strips, tiles, and blocks. Note the strip type of wood flooring in Fig. 26-21. Wood tiles are made up of small pieces of wood. See Fig. 26-22. Wood blocks are short pieces of wood. They are stood on end. The end grain wears better than face or edge grains.

Wood flooring is installed over concrete and wood subfloors. Concrete must be waterproofed. Fig. 26-23 shows two methods. Woodstrip floors are nailed to *sleepers* (strips of wood). Sleepers are held to the concrete with mastic, a thick glue.

Carpeting. Carpeting is placed over wood or concrete subfloors. Carpets often have pads under them. The pads are made of sponge rubber or plastic foam. A tack strip is placed around the edge of the room. See Fig. 26-24. The ***tack strip*** is made of wood and standard nails. Some tack strips are metal with hooks that hold the carpet edge. The pad is laid first. It is held in place with glue or staples. The carpet is cut to rough size. Next, it is rolled out in the room. Special tools are used to stretch it over the tack strip. Upended nails or hooks hold the carpet. Finally, the edges are trimmed and pushed below the level of the tack strip.

A

B

Fig. 26-20. Resilient flooring can be put over wood, concrete, or existing smooth flooring. A—Mastic being used to fasten it down over wood. (MacMillan Bloedel Ltd.) B—Adhesive is already on tiles. (Armstrong World Industries, Inc.)

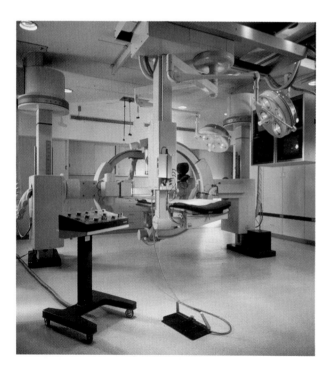

Fig. 26-19. A floor in a hospital must wear well and be easy to clean. This resilient flooring is comfortable to stand on. It gives a little when you step on it. (Armstrong World Industries, Inc.)

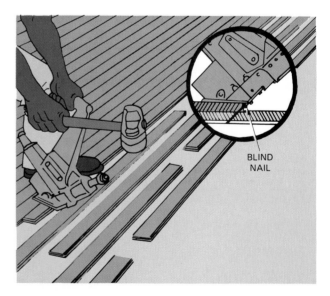

BLIND NAIL

Fig. 26-21. Wood strip flooring is nailed down. A special tool makes it easier. The angled nail is called a blind nail, since it will not be seen. (National Oak Flooring Manufacturers Assoc.)

Fig. 26-22. Wood tiles are laid one piece at a time. They are held in place with mastic. (Pease Flooring Co.)

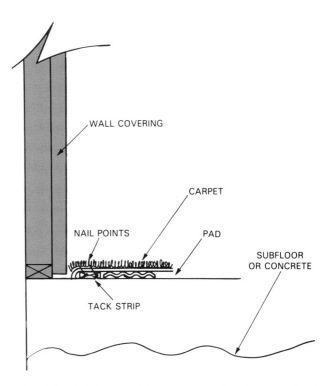

Fig. 26-24. Floors with carpets make the floor feel warmer. It is easier to hear in carpeted rooms.

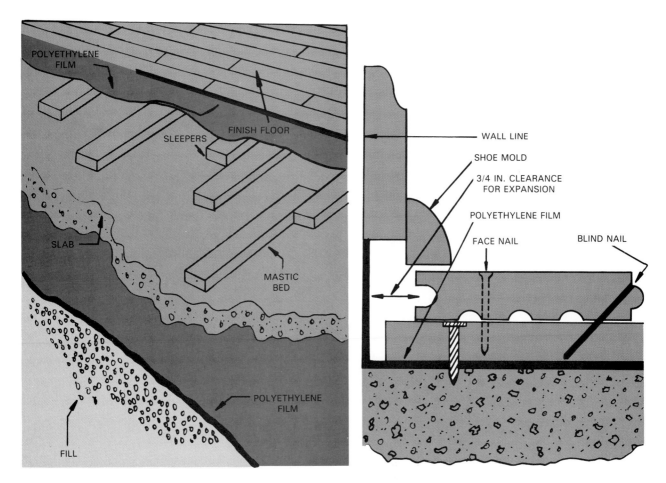

Fig. 26-23. Wood flooring must be kept dry. Plastic film keeps moisture away from wood. Two methods are shown. (National Oak Flooring Manufacturers Assoc.)

Trimming a Structure

Trimming is done to close up cracks, Fig. 26-25, and add beauty to raw, unfinished edges. Trim work involving moldings is done by the most skilled of carpenters.

Cracks appear when cover materials are not tight. Moldings and caulking are used to close cracks. Moldings can help attach windows or doors more solidly to framing.

Moldings

Moldings and trim are used on the inside and outside. Outside cracks are covered to keep out weather, birds, and insects. See Fig. 26-26.

Moldings are made of wood, plastic, and metal. Moldings like the one shown in Fig. 26-27 are made of wood and plastic. They are available as raw wood or prefinished. Raw wood needs to be painted or finished. That job is done before it is cut and nailed in place. *Prefinished molding* has a surface color or coating already on it. It only needs to be cut and nailed in place.

Special machines are used to cut exact angles on the ends. These machines make work easy and fast. See Fig. 26-28.

Metal, stone, and ceramic trim are also used. They are used more in commercial projects, such as at a corner between a floor and wall. Metal, stone, and ceramic trim do not require a finish. They are easier to clean and wear better than wood.

Caulking

Caulking is used inside and outside to keep water out. This helps prevent rot, crumbling, and paint peeling.

Most caulking is used to seal cracks around windows, bathtubs, and sinks. Caulking can be clear or can be opaque (blocks light), with several colors. Caulking comes in a tube with a built-in nozzle. The

Fig. 26-26. These trim boards conceal the rafters. Ventilators let air circulate in attic but keeps out animals and insects. (The Owens-Corning Fiberglas Corp.)

Fig. 26-27. This dark molding was prefinished. Prefinished means that it had the stain and coating on it when it was bought. (Western Wood Moulding and Millwork Producers)

Fig. 26-25. The wall covering does not meet the carpet. The base molding is used to trim the joint. It also protects the wall covering. Cleaning machines will not scuff it.

A

B

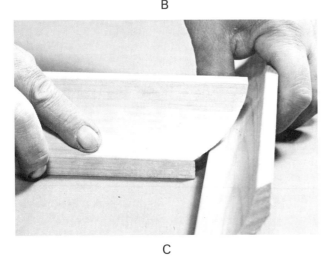

C

Fig. 26-28. Finish carpenters must be skilled. They use tools and machines to help them. A–Frame and trim saw. (Rockwell International Corp.) B–Miter saw being used to make 45 deg. cut on molding. C–Cove molding that has been cut with a coping saw, giving a coped joint. (Western Wood Moulding and Millwork Producers)

end of the nozzle is clipped off and the tube is put into a device called a gun, Fig. 26-29. The worker uses the gun to squeeze out a bead (a rope shape) of caulking. The nozzle forces it into the crack.

Silicone rubber is a good choice to seal cracks between tiles and fixtures. Some kinds are not used when a surface must be painted, because paint will not stick to them. Paintable silicone, latex, and oil-based caulking can be painted. The silicone and the latex cost more, but they last longer than oil-based caulking.

Fig. 26-29. Caulking is squeezed out of a tube with a gun. Both hand and air-powered guns are used.

Installing Equipment, Fixtures, and Furniture

Equipment and fixtures become a part of the structure. Often, finish work must wait until equipment and fixture connections are ready. The work on the walls and on the fixture cabinets must be done at about the same time. Furniture is not a part of the structure. Furniture installation does not interfere with wall and ceiling work.

Equipment

Equipment is installed throughout construction. See Fig. 26-30. Large generators in a power plant may be installed early. They would be too big to get into the building later.

Fixtures

Subcontractors return to complete the work on the mechanical systems. The elevator constructors put in the pushbuttons and lights, and finish the cars. The controls are connected and tested.

Fig. 26-30. Large pieces of equipment are installed before the structure is enclosed. (Southern California Edison Co.)

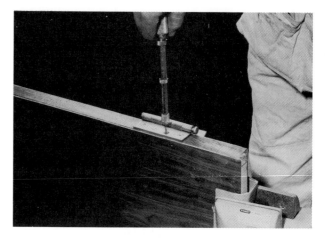

Fig. 26-32. This type of door is hung on the site. Note that a gain (a shallow recess for the hinge) is yet to be cut into the door stile (edge.) Some carpenters prefer to check the screw alignment first. (The Stanley Works)

Heating and cooling people install their units. They connect ducts and adjust the system. All parts of the system are tested.

Carpenters put in doors, stairways, cabinets, handrails, and curtain rods. Doors and stairways are installed before the painters begin. Interior doors may come in frames, Fig. 26-31. They are called *prehung doors.* With other doors, the frames, doors, and hinges are separate, Fig. 26-32.

Finish carpenters first set the frames. Fitting is done with shims. *Shims* are wedges of wood, usu-

ally sawn from cedar shingles. The frames are made plumb and square. See Fig. 26-33. The doors are hung on hinges and locks are then mounted.

Plumbers install plumbing fixtures like the dishwasher shown in Fig. 26-34. Plumbers connect ap-

Fig. 26-31. A prehung door is complete. It has a frame, hinges, lock, and a door. All are assembled at the factory. (Jordan Millwork)

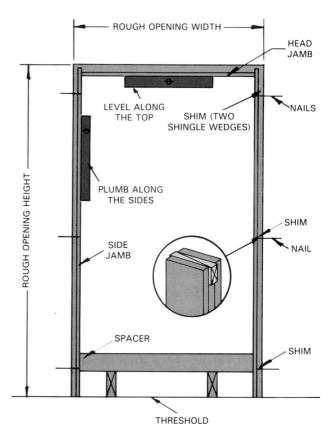

Fig. 26-33. Cedar shingles are used to plumb the door frame. The frame is held in place with nails. Once a pair of wedges is placed together, the excess is cut off flush with the door jamb.

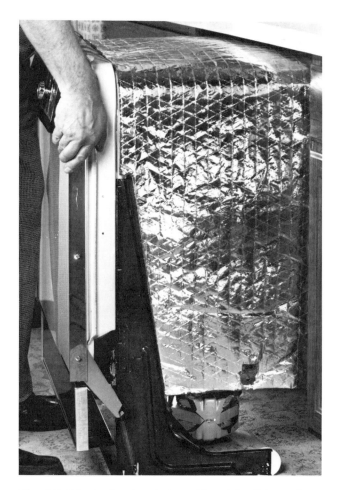

Fig. 26-34. A plumber is installing this dishwasher. Notice that the cabinets and resilient flooring have been installed. Note the insulation for sound reduction and for keeping heat in during the drying cycle. (Whirlpool Corp.)

Fig. 26-35. Receptacles are wired and covers are put in place after decorating is done. (Leviton Manufacturing Co., Inc.)

pliances and fixtures to the supply lines and drains. Piping systems are checked for leaks and flow rates.

Electricians install switches, outlets, and lights, Fig. 26-35. The supervising electrician checks out all circuits.

Communications technicians install the telephones and other communication devices. Units are connected and checked.

Furnishings

Furnishings include chairs, tables, and cabinets. Some structures have temporary wall systems. They are built in a factory. Parts are shipped to the site and then assembled. Wall surfaces and wiring are part of the wall.

If the walls do not reach the ceiling, they are called room dividers. See Fig. 26-36. Subcontractors often install these walls.

Note some differences between room dividers and temporary walls. Room dividers do not usually

contain wiring, but temporary walls can. Also, temporary walls can be fit to the floor and ceiling so well that they look like permanent walls.

Furniture is moved most easily with straps over your shoulders. This uses the natural leverage of the

Fig. 26-36. Engineered walls are easy to move. Room dividers are even easier. They do not reach the ceiling. Can you find two kinds in this picture?

body. Never lift furniture from a bent over position. Always stoop first, so the legs can do all the work in lifting.

SUMMARY

Finishing of structures includes painting, decorating, trimming, and installing. Finish work is done on the inside and the outside of the structure. Finish workers take special care in their work because their work is seen. Fixtures and equipment are installed at a convenient time for blending with the wall, floor, and ceiling coverings. Plans may indicate furniture placement.

KEY WORDS

All of the following words have been used in this chapter. Do you know their meaning?

Carpeting
Caulking
Decorating
Finish sheen
Grout
Interior designer
Lacquer
Painting
Pigment
Prefinished molding
Prehung doors
Resilient flooring
Shims
Sleepers
Stain
Tack strip
Terrazzo
Thinner
Trimming
Varnish
Vehicle

TEST YOUR KNOWLEDGE

Write your answers on a separate sheet of paper. Do not write in this book.
1. A structure is finished for three reasons. What are they?
2. Surfaces are made to clean easily and look good by _____ and _____.

3. Paint is made up of three ingredients. They are _____, _____, and _____.
4. An intermediate quality sheen on paint is called _____.
5. Three ways to apply paint are with _____, _____, and _____.
6. Three ways to decorate are with _____, _____, and _____.
7. It is easier to hang wallpaper if the paper is folded:
   a. Pattern side to pattern side.
   b. Paste side to paste side.
8. Describe terrazzo.
9. Describe resilient floor coverings.
10. In what three forms is it possible to buy wood flooring?
11. What materials are used to close up cracks when trimming out a structure?

Matching questions: On a separate sheet of paper, match the definition in the left-hand column with the correct term in the right-hand column.

12. _____ Helps make wood grain consistent.
13. _____ Will allow wood grain to show.
14. _____ Term for nail at 45 deg. in hardwood flooring.
15. _____ Material used to fill joints between ceramic wall tiles.
16. _____ Seals cracks around tubs, windows, and sinks.

a. Grout.
b. Lacquer and varnish.
c. Caulk.
d. Blind.
e. Stain.

ACTIVITIES

1. Compare your room with a classroom at school. List the materials, equipment, and furnishings for each. What different materials and methods were used to finish the two rooms?
2. Decide where you might add moldings to your room or an office. Make some samples showing joints for corners or where a molding overlaps an existing board. Use saws such as the miter saw. With your instructor's help, try common wood stains, varnishes, and lacquers to match or contrast two or more pieces.

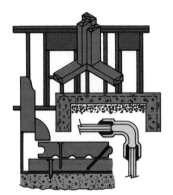

Chapter 27

LANDSCAPING AND OUTDOOR EQUIPMENT

After studying this chapter, you will be able to:
☐ Describe what landscaping is.
☐ Give examples of outdoor structures, equipment, and fixtures added to a building or road.
☐ Describe a landscape plan.
☐ Compare the uses for a landscape plan and for a site plan.
☐ List materials used for walkways, sidewalks, driveways, and platforms.
☐ Tell what must be done to the soil before planting grass seed or laying sod.
☐ Tell what is done to the soil before planting a tree or bush.
☐ Recall the proper way to put guy wires on a tree.
☐ List plants and materials used as ground cover.

During construction, supplies and waste collect on the site. See Fig. 27-1. The ground is rough and torn up. Large equipment, fences, and temporary buildings break up direct traffic routes. This all changes when the structure is nearly complete.

Landscaping can begin when equipment and material storage needs allow for it. There are many jobs that are a part of landscaping work.
• The holes are filled.
• The piles of material are used or hauled away.
• The debris is cleaned up.
• Large machines are moved to another project.
• Temporary fences are taken down.
• Temporary buildings are removed.
• The site is leveled or contoured.

Fig. 27-1. During construction the site is cluttered.

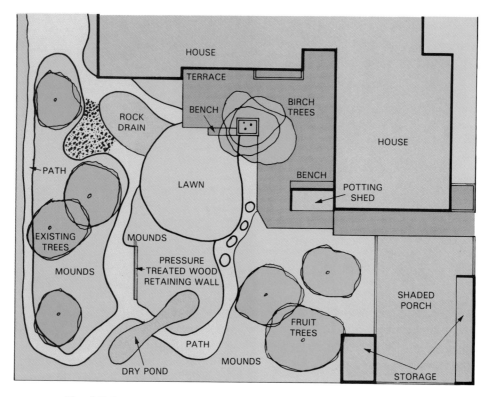

Fig. 27-2. The landscape plan is a drawing of the finished site.

- Sidewalks are poured.
- Trees, shrubs, and grass are planted.
- Fixtures (outdoor equipment) are placed.

This chapter discusses five processes. They are: doing final earthwork, building accesses, installing fixtures, planting the site, and cleaning the site.

WHO DOES SITE WORK

The owner and landscape architect work with the head architect. They produce the *landscape plan.* See Fig. 27-2. The plan locates all structures and features. It describes the shape of the finished site. The location of each planting, walkway, or special piece of equipment is shown.

A high-rise building may need little site work done, Fig. 27-3. A golf course, Fig. 27-4, is almost all site work.

Some plans are small and simple. See Fig. 27-5. Others are large and complex. See Fig. 27-6.

Earthwork is usually done by general contractors. They backfill around the foundation of structures. They then build the foundations for fixtures and construct walkways.

The landscape subcontractors often provide the plants and fixtures. They finish the grading and cultivate the soil. Planting, watering, and cleanup are their jobs, too.

Fig. 27-3. The site work is a small fraction of the total cost for this structure. There is little land for trees, plantings, and walkways.
(The Stubbins Associates, Inc.)

DOING THE FINAL EARTHWORK

As noted before, the earth is cluttered during construction. There are holes, trenches, scraps of material, and piles of dirt.

Earth was piled for later use during final earthwork. In contrast to early earthwork, final earthwork involves replacing and shaping the earth mainly for appearance rather than for support.

Fig. 27-4. Site work is the major building cost for a golf course. (Robert Trent Jones II)

Fig. 27-5. This landscape project was simple. A little grading and a few plantings were all that was needed. (California Redwood Assoc.)

Fig. 27-6. Much planning, work, and material went into this landscape project. (Pella/Rolscreen Co.)

Replacing Earth

Holes may have been dug for basement foundations, pipelines, and underground tanks. The soil that was saved is used to backfill, Fig. 27-7. Backfilling fills in around structures above ground or those below ground like pipelines. In the process, the soil is compacted. Air is squeezed out, making the soil firm so it will not settle and crack foundations or ruin landscaping.

Topsoil is the best soil for growing plants. It is scraped from the top one or two feet of the site. Topsoil is piled out of the way and saved. It is spread over the site after the site is leveled. Some sites do not have good topsoil. It has to be hauled from another site.

Shaping Earth

The site can be shaped to make it like a forest, a meadow, a valley, a wild area, a garden, a clearing, or like a "business park." See Fig. 27-8. Extra soil can be reshaped into mounds. The shape can provide drainage, add beauty, or make the changes in elevation (height) gradual and easy for walking.

The site plan describes the shape of the site. It shows where mounds and flat areas are placed. The elevation and shape of each is shown. Some machines used to shape sites are shown in Fig. 27-9. Bulldozers and graders move, level, and mix soil. Loaders and scrapers move soil to where it is needed or remove excess soil. Small jobs are done with shovels, rakes, and wheelbarrows.

Fig. 27-7. The fill dirt on the left will be used to cover a sewer line. This is called backfilling. The backhoe will be used to do the work.

Fig. 27-8. This site was shaped. Notice how walls allow the steps to be on gentle slopes. (California Redwood Assoc.)

BUILDING ACCESSES

Buildings need driveways, loading docks, parking areas, and walkways. These are called *accesses.*

A

B

Fig. 27-9. These machines move, level, mix, and compact soil. A–Large trucks can move soil quickly. (Caterpillar Inc.) B–A roadside is finished by grading and smoothing.
(Puckett Brothers Manufacturing Co., Inc.)

Driveways

Driveways are extensions of a street onto a site. See Fig. 27-10. Like streets, they are designed for the type of traffic. The strength of driveways or roadways depends on three things:

1. The volume of traffic. The more vehicles that travel it, the stronger it is built.
2. The maximum weight of the vehicles. The larger the load, the stronger the road is built.
3. The speed the vehicles travel. Fast traffic can create ripples in a road surface after long use. Stronger roads are needed for faster traffic.

Loading Docks

Plans will include loading and unloading access at industrial sites. Most factories need *loading docks* for semitrailers. Other factories need rail and car access. Some need shipdocks.

Walkways

Walkways are made in many shapes and with many materials. Concrete walkways are the most common, Fig. 27-11. They last the longest and are

Fig. 27-10. A driveway leads to a parking lot for this sports arena. The walkway and entrance invite people to enter. (TEMCOR)

Fig. 27-11. These walkways are made of concrete. Notice how walks keep soil types separate for easy filling. (DYK Prestressed Tanks, Inc.)

easy to build. Concrete can be cast in any shape and finished in many ways.

The bearing surface for concrete walkways and roadways is compacted soil. Porous (lets water through) gravel fill is the foundation. Steel rods and wire are used to reinforce concrete. The surface is finished with brooms, trowels, or texturing tools.

Pressure-treated wood, stone, brick, and asphalt are also used. More patterns can be made with some of these than with concrete.

Pressure-treated wood lasts long and does not rot quickly. It makes a walkway with good drainage, Fig. 27-12. Wood walkways are nailed to stringers. ***Stringers*** hold the top boards together and off the ground. They run the length of the walkway. One is placed on each side. They are laid in a bed of gravel or attached to posts or concrete footings. The boards on the walkway surface are nailed to each stringer, Fig. 27-13.

Brick and stone are set in a bed of sand or on a concrete slab. Sand or mortar is used to fill in the spaces between units. An example of a brick walkway is shown in Fig. 27-14.

INSTALLING OR BUILDING FIXTURES

Fixtures are then added to the site. Fixtures are small structures. They include lights, signs, railings, seats, shelters, and other items not considered part of the main structure. Some of them have substructures and superstructures.

Some fixtures require electric or gas lines. The size of foundation varies. It depends on the bearing surface, frostline, and forces on the superstructure.

The superstructures are often manufactured. See Fig. 27-15. They are installed by the general con-

Fig. 27-12. Wood walkways are attractive. The lines show where stringers are placed.
(Product of Wolmanized® Pressure-Treated Wood)

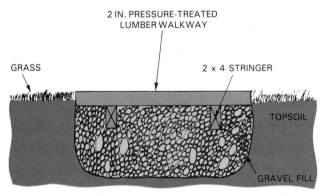

Fig. 27-13. Pressure-treated wood walkways are easy to build.

Fig. 27-14. Brick and stone make good walkways and driveways. They are made in many colors, styles, and textures.

Fig. 27-15. Note the fixture that was manufactured on this site. (Scyma Div. Michigan Industrial Co., Inc.)

Fig. 27-16. Common materials were used to build these fixtures on the site.

Fig. 27-17. The plants and shape of the site were left natural in this landscape plan. (Pella/Rolscreen Co.)

tractor or the landscape subcontractor. Others are built on the site, Fig. 27-16.

A *sprinkler system* is a fixture used to water grass and flowers. Water pipes are run underground. Sprinkler heads are placed so that the entire site is watered. The system can be turned on by hand or by timers. With some systems, the heads of the sprinklers drop below the grass when the water to the system is turned off.

Fixtures on highway projects include railings, lights, striping, signs, and reflectors. Most of them are built to increase safety.

PLANTING THE SITE

Planting makes the site and structure look natural. The land around some projects is already beautiful. The architect may design the building to fit the natural setting. See Fig. 27-17. If the land is plain, it may be changed with ponds or stone walls, Fig. 27-18. Note in the two illustrations that more landscape work was done around the second building than the first.

Existing plants can be left in or new plants and grass put in. Steps for putting in plants include cultivating, fertilizing, digging, and planting.

Preparing the Soil

The soil must be prepared for planting. Rototillers or harrows are used. They have fingers that dig into the soil. The clods are broken and rocks are brought to the surface for removal. This work is called *cultivating.*

Fertilizer is spread and mixed with the soil to make the plants grow better. Other chemicals are used to destroy harmful insects and weeds.

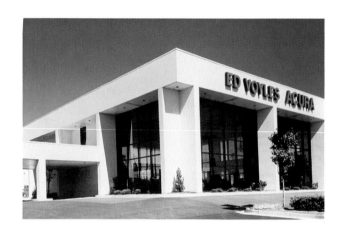

Fig. 27-18. This building has simple lines. The landscape is free of clutter, too. (Stran Buildings)

Planting Trees, Shrubs, and Flowers

The landscape plan describes the kind and placement of each tree, shrub, or flower area. See Fig. 27-19. When timing is important, a *planting schedule* is used. Digging the proper sized hole for a tree is done before putting in a lawn.

The bottom third of a planting hole is filled with water to keep air from the roots and to soften the soil. The tree, shrub, or flower is placed in the hole, Fig. 27-20. Soil is replaced and packed around the roots. No air pockets should remain around roots. More water may be added.

Trees need extra support until the roots grow. Three or more ropes or wires and some padding are used. These lines, sometimes called *guy wires,* are tied to the tree trunk, Fig. 27-21. Stakes hold the

Fig. 27-19. The landscape subcontractor follows a plan. The placement of trees, shrubs, and rocks is shown.

other end of the support line. Each support line goes in a different direction. Trees and plants are watered at least every two days until their roots are established.

Using Ground Cover

Ground cover is used to keep the soil from washing away. It can also cover unattractive soil. Ground cover usually means some kind of low plant, but the term can refer to wood and mineral products used for mulch. *Mulch* is a covering of small wood chunks or other material over the soil or mixed with the soil. Bark and chipped branches are common. See Fig. 27-22. Rock, chipped marble, and crushed brick last longer.

Fig. 27-20. Trees can be transplanted (moved) from one place to another. (Vermeer Manufacturing Co.)

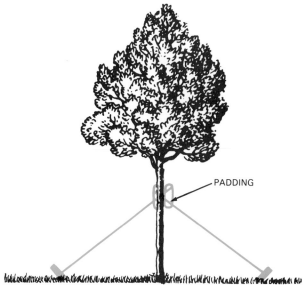

PADDING

Fig. 27-21. Padding protects the tree trunk from damage.

Fig. 27-22. Grass and coarse bark mulch are used for ground cover at this home. Bark discourages weeds. The driveway has good drainage with gravel. (Pella/Rolscreen Co.)

The most common ground cover is grass. Lawns are started by planting seed or placing sod. *Seeding* a lawn costs the least, but the grass takes more time to grow. Placing sod, or *sodding,* is faster but more expensive.

Seed is spread by hand or machine, Fig. 27-23. Rakes are used to mix the soil and seed. The seedbed is rolled. Seeds grow better if they are tight against the soil. An upside-down leaf rake can help push seeds into the earth. The soil is covered to help it hold moisture. Straw or burlap is used. See Fig. 27-24. It also keeps the soil from washing and blowing away. The soil is kept moist until the lawn gets a good start.

Fig. 27-23. A hydro-seeding method is being used. Water, fertilizer, binders, seed, and mulch are being applied. The mulch protects the seeds from birds and holds moisture. (Veasey Seeding, Inc.)

Fig. 27-24. A seeded patch. Cover freshly seeded areas with straw or burlap to hold in moisture and to discourage birds. (California Redwood Assoc.)

Fig. 27-25. What method was used to start this grass? (Manville Building Materials)

Sod is laid over the bare soil. See Fig. 27-25. It is tamped in place and watered. The sod is watered twice a day until the roots are well established. In some areas where it is dry, even mature lawns are watered during the entire growing season.

CLEANING THE SITE

Cleaning involves picking up and disposing of all debris. Debris includes empty bottles, cans, boxes, and scrap of all kinds. Fixtures must be cleaned and walkways swept. Cleaning includes removing any straw or burlap from a lawn project once the lawn is established. Cleaning can also involve removing stains on building materials. A weak hydrochloric acid solution is often used to remove rust stains.

SUMMARY

Landscaping improves drainage and makes the site look better. Architects work with the owners and landscape architects to draw up landscape plans. General and subcontractors carry out the plan. Landscaping involves earthwork, and building accesses and fixtures, in addition to planting and cleaning.

KEY WORDS

All the following words have been used in this chapter. Do you know their meaning?

Accesses
Cleaning
Cultivating
Driveway
Fertilizer
Ground cover
Guy wire
Landscape plan
Loading dock
Mulch
Planting schedule
Pressure-treated wood
Seeding
Sodding
Sprinkler system
Stringer
Topsoil
Walkway

TEST YOUR KNOWLEDGE

Write your answers on a separate sheet of paper. Do not write in this book.
1. Name at least three things that are described on the landscape plan.
2. Who usually does most of the earthwork?
3. Who provides the plants and does the final grading, planting, and the final cleanup?
4. Final earthwork is done mainly for _____ rather than for _____.
5. The strength needed for an access roadway depends upon traffic _____, _____, and _____.
6. Name two building accesses needed by a home.
7. Name four materials used to build walkways.
8. The boards on a walkway surface are nailed down to:
   a. Struts.
   b. Joists.
   c. Pilasters.
   d. Stringers.
   e. Walers.
9. The steps for putting plants on a site include _____, _____, _____, and _____.
10. Grass seeds are pushed into the earth with an upside-down leaf rake or with a _____.

ACTIVITIES

1. Ask a landscape architect to your class. Have him or her describe how to plan projects.
2. Make a sketch of your backyard. Imagine improvements you could make. Develop a landscape plan. Call a nursery to find out what it would cost to carry out the plan.
3. Visit a construction site that is being landscaped. Make a list of machines and tools being used. What are workers doing? What work would you choose to do?

Section 8
CLOSING THE CONTRACT

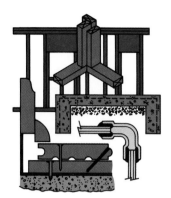

Chapter 28

TRANSFERRING THE PROJECT

After studying this chapter, you will be able to:
☐ List three things that an inspection involves.
☐ Name the individuals and commissions that make inspections.
☐ Recall common construction defects that might require correction.
☐ Recall steps for resolving financial claims.
☐ List names of documents involved with financial claims.

When all work is complete the project is transferred. Responsibility changes from the contractor to the owner. The contract is the guide in this action. A contract sets up six conditions for a successful project:
• It states the specifications for the project.
• It describes the materials, methods, and quality of work.
• It spells out the process used to inspect and correct defects.
• Standards for paying bills are stated.
• Warranties are described.
• It tells how final payment is to be made.

MAKING INSPECTIONS

Inspections are made throughout construction of a project. Officials started by looking over the set of plans. The plans had to be approved before a building permit was issued. Inspection continued throughout construction to see that plans and specifications were followed. See Fig. 28-1. A final inspection is done before transferring the project.

Inspections never completely end. The final inspection only relates to building the project. Public health, fire, and insurance inspections continue. They go on during the entire lifetime of a structure. See Fig. 28-2.

Fig. 28-1. Work is inspected as it is done. (Honeywell, Inc.)

Fig. 28-2. Fire inspectors check the condition of a structure. They see that exits are large enough and are locked only from the outside.

Who Inspects

There are seven main groups of people who make inspections:

- Workers.
- Contractors.
- Public officials.
- Project designers.
- Lenders.
- Owners.
- Suppliers.

Inspectors must be qualified. Fig. 28-3 shows an inspector at a site. *Qualified inspectors* help others do their jobs. A person inspecting plumbing must know about plumbing. This means the person must know the right order for connecting all pipes and fittings.

Workers are responsible for finding defects early. They check their work and others. They take pride in their work because their future employment depends on it.

Contractors must satisfy a contract, Fig. 28-4. They will not get paid until they do. It is cheaper to do a job right the first time. If errors are made, contractors want to find them early. Leaky pipes are easier to fix before they are covered. Future contracts depend on past work. For this reason, contractors try to do their best work on time with few problems.

Public officials are concerned for public safety, Fig. 28-5. They see that the structure is strong enough. The materials, size, and methods are checked by public officials. They compare what was done with what is required by the building code. The project must meet or exceed the requirements.

Project designers inspect to see that the design was carried out. A glass wall (made of glass block)

Fig. 28-4. The contractor (middle) is checking on a step in the construction process. The owner (right), architect, engineer, and supervisor (left) are consulted. Good communication helps the contractor build the structure correctly.
(Southern Company Services)

Fig. 28-3. Work must comply with construction documents. Here a certified inspector checks the placement of rebar. (Western Technologies, Inc.)

Fig. 28-5. Building inspectors protect the public. If a part of a structure is not right, the contractor must correct the problem. (U.S. Elevator)

that leaks will not pass an architect's inspection. Pipelines must move enough product before engineers accept them.

Lenders, the people who finance the project, inspect the project, too. To be a good investment, the project must work well. Lenders do not release partial payments until they are satisfied with the materials and labor used.

Owners inspect the project or hire someone to do it. Architects and engineers often represent the owner. Owners usually have a better idea of the expected results than other groups.

Suppliers check that shipments of their products are what the customer ordered. It is costly to deliver bricks that are the wrong shade. The shipment must be picked up and replaced with another. This causes delays in the schedules for both the supplier and contractor.

What is Inspected

Inspectors look for three main things. They check the materials, methods, and quality of work.

Materials

Before building permits are issued, materials are checked on the plans. Public officials see if they meet the code. The size and kind must be right.

After plans are approved, they become the guide. Inspectors must see that the specified materials are used. This must be done during construction. Tests are run to find out if materials are strong enough. See Fig. 28-6. Concrete is tested. Each batch of concrete is sampled on some jobs.

Nearly all work is covered when the structure is enclosed. The final inspectors check the surface materials. At this time, trim, paint, plantings, and fixtures are checked against the plans.

Methods

Some methods for a job are better than others, Fig. 28-7. Designers cut costs with the choice of method. The drawings show the materials and where they are placed. Specifications state how work is to be done. For example, the methods used to backfill are specified. In general, backfill is to be placed evenly to avoid uneven soil pressures. Topsoil, 4 in. deep, is used in lawns.

Quality

The quality is checked in several ways. Inspectors answer the following questions:
• Does it work?

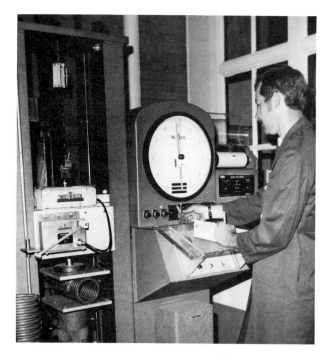

Fig. 28-6. These sections of plastic drain pipe must support soil when pipe is buried. Samples are tested for crushing strength. (Pennsylvania Department of Transportation)

Fig. 28-7. A technician checks the quality of full penetration welds using ultrasonic methods. (Western Technologies, Inc.)

- Is it the right size?
- Does it look good?

Most often the first question deals with mechanical systems. Heaters, lights, sinks, and drains must work as planned. Pipes must withstand the working pressure. Circuits must be able to carry the specified electrical load.

The size of the structure must be right. Walls, rooms, steps, and grades must be the right length, width, and height. Inspectors check the size and placement of openings and fixtures. See Fig. 28-8.

Fig. 28-8. Inspectors make sure openings are the right size and place. Blocklaying and grouting are inspected. Thickness, height, and defects in parging and tar coating are checked.

High quality work looks good. Mortar joints must be neat and uniform. Floors, walls, ceilings, and fixtures should not be scratched or marred. Curves in sidewalks and flower bed edges should be smooth. Trees must be straight and sod must be even.

WHAT A FINAL INSPECTION IS

Five steps are involved with inspections:
- Find a standard.
- Make an inspection.
- Make up a punch list.
- Make corrections.
- Sign approval forms.

The contract states who must approve the project. The owner, architect/engineer, and public officials are involved. In the final inspection, materials and quality are checked and inspectors try to learn what methods were used, Fig. 28-9. Inspectors begin by finding inspection standards in the contract and in drawings. Defects are placed on a list called a *punch list.* Each item on the punch list must be corrected by the responsible firm. This firm may be the general contractor, subcontractor, or supplier. Corrected items are approved. *Approval forms* are signed by those who made the final inspection.

Fig. 28-9. A technician uses a calibrated torque wrench to confirm tension of high-strength bolts. (Western Technologies, Inc.)

CLOSING THE CONTRACT

At the *contract closing,* the owner pays the contractor and receives documents. The details of the closing are written in the contract. The contractor must provide signed approvals, releases, warranties, and manuals. The owner makes the final payment to the contractor.

Approvals

A *certificate of substantial completion* is issued when the project is nearly complete. Only final cleanup, some painting, etc., may remain. At this time the final inspection is made. The final inspection often concentrates on the appearance of the project and on mechanical equipment. The contractor, subcontractor, and supplier correct the defects caused by anything other than designer error or the owner's poor planning. Some rework is done at the cost of the owner if necessary. The inspectors are brought back to approve the work. When work is approved, a *certificate of completion* form is signed. The paper states the dates for the warranty. A warranty is a promise to repair things that result from poor work.

Releases

In construction, a contract means that people promise to do things for a fee. The contract describes the terms of the promise and the fee. For example, contractors promise to pay for materials, labor, and equipment. Sometimes these payments are not made. In that event, legal action is taken. A payment schedule is set up. Sometimes a compromise is made. When the problem is resolved, a release is signed. A *release* frees a person of the immediate complaint, sometimes in exchange for a long-term responsibility.

Complaints of nonpayment are of two types. They are called the claim and lien. Fig. 28-10 is a flowchart that includes steps showing how to resolve complaints of nonpayment.

Claims

Claims arise when extra work has to be done. Let us say a contractor is building a house. The plans did not show a buried dump site. The old trash had to be excavated and hauled away. It added a cost to haul and dump the trash at the landfill, and five extra loads of fill gravel had to be used. A deeper foundation was required. The contractor feels the owner should pay the extra expense.

The first step is to settle through negotiation. The owner and designers talk with the contractor. If the problem is solved, a *release of claim* is signed. Legal action is taken if a solution was not found.

Legal action takes place in a court of law. Each party's views are stated. A judge decides the issue. If the contractor wins, the owner must pay the extra costs. After payment is made, a release of claim is signed.

Liens

Liens arise when an owner owes money on the property. This can be from an old debt or it may be owed to workers, subcontractors, or suppliers. For example, money can be withheld because the owner is not happy with work or materials and the owner does not pay the contractor. As a result, the contractor does not pay either the person who did the work, or the supplier.

Legal action is taken by the person who has not been paid. If the owner is right, the person does not get paid. If the owner is wrong, a *mechanic's lien* is given. A person with a mechanic's lien can force the sale of the property. The money received for the sale pays the person who holds the lien. The remainder goes to the owner.

When the project is complete the owner gets a *release of lien.* This tells the owner that nothing is owed against the project.

Warranties and Manuals

A *warranty* is a document giving a guarantee that there are no defects. There are two kinds of warranties: from the contractor and from the supplier. The contractor states that the work has no defects. The supplier says there are no defects in materials.

The warranty statement applies for a stated length of time. The length of time is a part of the certificate of completion. Roof material and work may carry a 20 year warranty. Most warranties are limited to one year.

Another item needed by a new owner is a manual. Equipment suppliers provide manuals that describe how to operate and maintain the equipment. Parts lists are often included. All manuals are turned over to the owner at the closing of the contract.

FINAL PAYMENT

The contract is completed when:
- Work has been finished.
- Final review is completed.

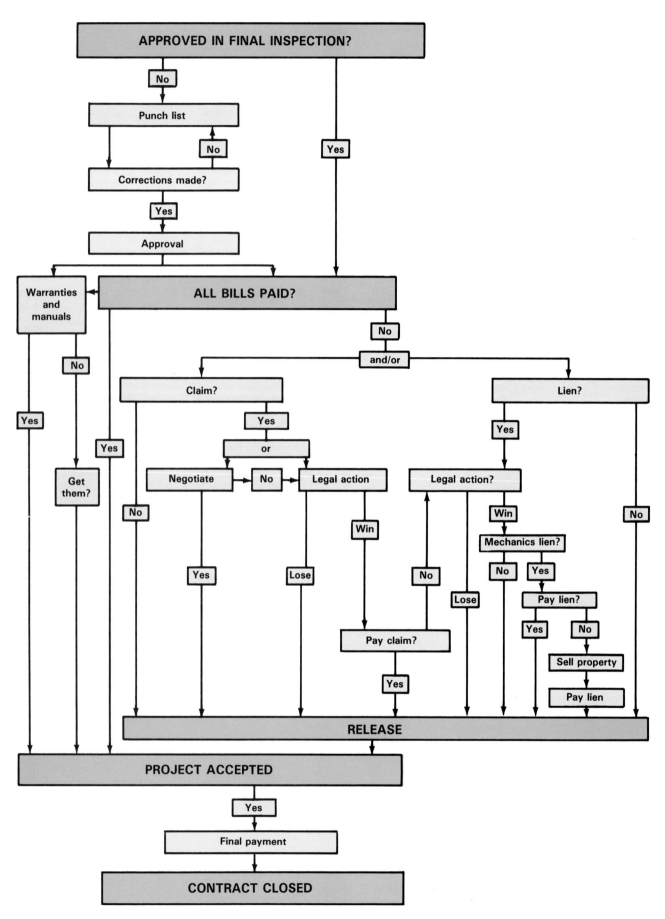

Fig. 28-10. The procedure for closing a contract is shown in this flowchart. The procedure moves from the top to the bottom.

- Releases are signed.
- Warranties and manuals are turned over.
- The certificate of completion is signed.

Now the owner makes the final payment. The contractor is released from the performance and payment bonds. Bonding is described in Chapter 9. The contractor's responsibility ends except for the warranties. The contract is closed.

SUMMARY

Inspections begin when the building permit is issued. They end when the structure is taken out of service. Inspectors are concerned about materials, methods, and quality. Near the end of construction, a certificate of substantial completion is signed. The final inspection is one that concentrates on appearance and/or mechanical equipment. When no defects are found a certificate of completion is signed. The owner must be assured there are no claims or liens on the project. A release of all claims and liens frees the owner of this responsibility. Warranties and manuals are given to the owner. An acceptance form is signed, final payment is made, and the contract is closed.

KEY WORDS

All of the following words have been used in this chapter. Do you know their meaning?

Approval form
Certificate of completion
Certificate of substantial completion
Claim
Contract closing
Legal action
Lien
Mechanic's lien
Punch list
Qualified inspector
Release
Release of claim
Release of lien
Warranty

TEST YOUR KNOWLEDGE

Write your answers on a separate sheet of paper. Do not write in this book.

1. Before getting paid, a contractor must satisfy the _____.
2. Public officials are concerned about public _____.
3. Inspectors examine _____, _____, and _____.
4. Which of the following conditions is not checked while checking quality of work?
   a. The mechanical systems work.
   b. The materials are those called for in the plan.
   c. The structure is the right size.
   d. The structure looks good.
5. When the project is nearly complete, a certificate of _____ _____ is issued.
6. During the final inspection all defects are placed on a _____ list.
7. When a problem in payment is resolved, a _____ is signed.
8. When extra work has to be done due to an oversight by the owner, the contractor has a _____.
9. When the owner owes money on the property, a _____ is put on the property.
10. A _____ is a document guaranteeing that there are no defects.

ACTIVITIES

1. Think of the last thing you purchased. Relate what you have learned to that purchase. Did you have a standard in mind for the number of defects you would accept? Were you prepared to ask for a price reduction?
2. Talk with the owner(s) of a new structure. Have them explain their contract closing procedure. How long do warranties last? How long can a payment schedule term last?
3. Look at the owner's manual for a home appliance. Also get a service manual for a device. Does each have a parts list?

Restoration is a way of changing a project. It saves money and materials and preserves our culture. (BIL-JAX, Inc.)

Chapter 29

OPERATING, MAINTAINING, AND CHANGING THE PROJECT

After studying this chapter, you will be able to:
☐ Plan for or schedule servicing work.
☐ Take advantage of a warranty when a project develops a defect.
☐ Discuss remodeling when bearing walls are present.

The owner now has the responsibility for the project. *Operating* a project can include both maintaining it and using it. Some projects are complex and hard to operate. See Fig. 29-1. Other projects are simple. See Fig. 29-2. An owner must *maintain,*

Fig. 29-2. The circle is the outside of a large buried tank. The tank and tennis courts need little maintenance. (DYK Prestressed Tanks, Inc.)

(repair and protect) the project. Sometimes, if a warranty is still in effect, the owner does not need to pay all of the cost of upkeep. The warranty requires the builder or supplier to fix a problem free of charge.

To increase a project's value and use, it is sometimes changed. The owner can change it in three ways. *Changing* includes altering or restoring, or installing new equipment in a structure.

OPERATING THE PROJECT

Contracts may include training for the operators of a project. Operators can check the equipment using dials and gauges that show how well machines are running. See Fig. 29-3. Climate control equip-

Fig. 29-1. This is the control room of an electric power generating plant. Gauges show if machine components are in good condition.
(Tennessee Valley Authority)

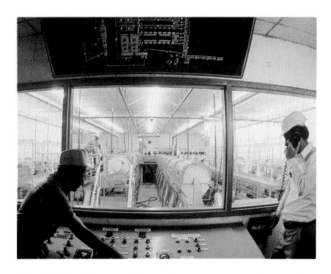

Fig. 29-3. These dials and gauges monitor equipment operation. They tell the workers if machine movements are within a standard. They give warnings when things go wrong. (Bob Dale)

ment is checked this way. More information comes from scheduled inspections. Overheated motors, worn pumps, and bare wires may be found this way.

Occupants and workers can report problems. Leaks are found by people who work on equipment or live and work near the plant.

MAINTAINING THE PROJECT

Maintenance extends the life of the structure. Well-maintained structures last longer. When maintenance is scheduled, it is called *scheduled maintenance.* Some schedules specify work on a daily basis. Banks and schools are cleaned daily. Windows may be washed once every six months.

Friction is the main problem with machines. Oiling and greasing a machine or fixture will reduce friction and reduce wear on its parts, Fig. 29-4. This maintenance work is also scheduled. It is called *preventive maintenance.*

Another example of scheduled work is street sweeping, Fig. 29-5. This schedule depends on many things.

Some maintenance is done when it is needed. This is called *routine maintenance.* Two examples are cleaning roots out of a sewer line and pruning trees, Fig. 29-6.

Repairing Projects

Parts of projects that fail must be repaired. See Figs. 29-7 and 29-8. Light switches wear out. Drains corrode.

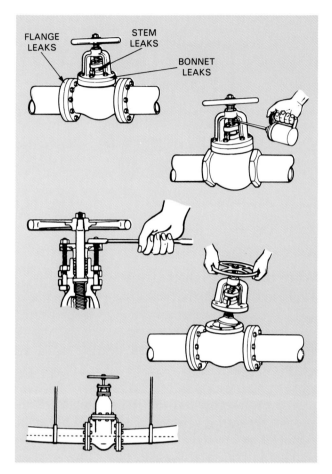

Fig. 29-4. Maintenance of plumbing equipment done by either the builder or the owner. Oil the valve stem regularly. Inspect disk assembly to find problems early. Realign pipe hangers just after initial installation. (Crane Co.)

Fig. 29-5. Examples of scheduled maintenance. Streets are cleaned two to three times per year. A similar scheduled job involves cleaning windows like those in the background. (Elgin Sweeper Co.)

Fig. 29-6. Ice storms damage trees and sewer lines. This kind of maintenance is not scheduled (Asplundh)

A

B

The roof may begin to leak. See Fig. 29-9. Flat roofs must be sealed better than pitched (slanted) roofs. Driveways need repair after many heavy loads are hauled on them. A careless worker may damage a structure, or a project may be getting old. All of these conditions will require repairs.

When equipment fails, it is best to contact the supplier or company that installed it. Often, the knowledge or tools for repairing it are not available to the owner.

Repairs should be done right away. If delayed, the problem may become worse. Structures that are not repaired lose value.

Protecting the Project

Projects must be protected from damage, theft, and fire. Protection involves three things:
- Fences.
- Patrols.
- Monitoring devices.

Fences are a low cost form of protection. A fence around a swimming pool helps keep it safe. Many cities require a pool fence to be at least four feet

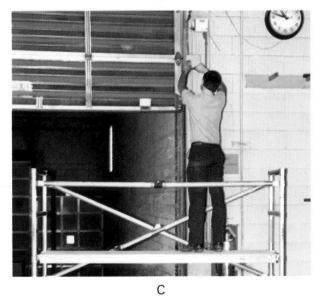

C

Fig. 29-7. Examples of typical repair work. A—Highways need continual upkeep. (Ingersoll-Rand Co.) B—Pipelines corrode. (Colonial Pipeline Co.) C—Adjusting an overhead door track. (BIL-JAX, Inc.)

A

B

C

Fig. 29-8. Much repair work consists of replacing parts. A–Servicing a pneumatic tube system. (Translogic Corp.) B–Replacing a section of sanitary sewer. (American Cast Iron Pipe Co.) C–Replacing an old telephone pole. (Bob Dale)

Fig. 29-9. Roof repair is the most common servicing task on any building project. (BIL-JAX, Inc.)

high. Factories with a lot of equipment that is not enclosed need fences. See Fig. 29-10.

Patrols watch property to keep it secure. They watch for people who might cause a problem. Patrol persons help direct people and keep them out of hazardous or restricted areas. Airplanes are often used to patrol pipelines, highways, and power

Fig. 29-10. Painting and rust prevention are major tasks to maintain this plant. Fences and patrols are another requirement for operating a project. (Libbey-Owens-Ford Co.)

transmission lines. The airplane in Fig. 29-11 is patrolling a pipeline.

Monitoring devices help to protect, too. A closed circuit television system can monitor safely an entire factory.

Monitoring systems such as smoke alarms can warn people. They can be tied into the city fire alarm system. If a fire occurs, workers, fire fighters, and others are warned. Sprinkler systems start automatically if the area gets too hot. People in many structures have planned ways to respond to fires.

CHANGING THE PROJECT

To change a project means to alter the structure, install equipment, or restore the structure. The change is planned, built, and serviced just like a new structure.

Altering the Project

Altering a structure means to change its size or shape. It is sometimes referred to as *remodeling.* The change can occur inside or outside. A structure is often altered when the use of the structure changes. The use of the structure in Fig. 29-12 was changed.

You alter the outside structure when you build an addition, Fig. 29-13. An addition raises the value of a structure and expands its usefulness.

You can also alter the outside of a structure by changing how a structure looks. This change may include adding or taking out windows. Changing rooflines is another way to alter a structure.

When the use of a structure changes, rooms on the inside are often changed. Rooms are altered by adding and removing walls. Care is taken when removing walls. Some walls, called bearing walls, support the floors or roof above. A plan for remodeling must consider all effects on the existing equipment and structure.

Fig. 29-12. This old mansion was restored and changed into a conference center.

Fig. 29-11. People in the airplane are patrolling the pipeline below. They are checking for trespassers and maintenance problems.
(American Petroleum Institute)

Fig. 29-13. More space is being added to this structure. (BIL-JAX, Inc.)

Fig. 29-14. These computers are used to monitor the operation of an electric power generating plant. Every so often new ones are installed to replace worn-out or out-of-date units.

Installing

Installing means to place something new in a structure. It can be a furnishing, such as new carpet, or equipment. Replacing old equipment such as a washer, or adding a sprinkler system to lower insurance costs, are other types of installations.

Computers are installed in some large structures, Fig. 29-14. They control lights, heat, ventilation, and cooling. They are called energy management systems. They pay for themselves with the money they save. This is called having a *payback*. New furnaces, air conditioners, and water heaters also have a payback.

Restoring Structures

Some people feel older structures should be saved. These structures are symbols of the past. They represent how people used to live. If they were lost, a part of our culture would be lost. For these reasons, people restore some structures. The structure in Fig. 29-15 was restored.

People who *restore* structures study how structures were built in the past. Then, they try to copy the techniques used. The restored structure looks much like it did when it was new.

Fig. 29-15. Restoring a structure is a good practice. It saves money and materials. It also preserves our culture. The illustration shows the structure's condition before and after remodeling. (©Cardinal Industries, Inc.)

SUMMARY

After the contract is closed the owner must service the project. Owners use and sometimes change the project. They must maintain, repair, and protect the project. A project is changed when one alters or restores it, or installs equipment. Some new projects or pieces of equipment are planned for a specified payback.

KEY WORDS

All of the following words have been used in this chapter. Do you know their meaning?

Altering
Changing
Fence
Installing
Maintaining
Monitoring device
Operating
Patrol
Payback
Preventative maintenance
Remodeling
Restoring
Routine maintenance
Scheduled maintenance

TEST YOUR KNOWLEDGE

Write your answers on a separate sheet of paper. Do not write in this book.

1. Projects last longer when they are _____.
2. When parts of a project fail they must be _____.
3. A piece of property and the people who work there are protected by _____, _____, and _____ _____.
4. When you want to change the _____ and _____ of a structure, you alter it.
5. Describe how a structure is restored.

ACTIVITIES

1. List and describe how your school is maintained. Who maintains and repairs it? Find out if any defects were repaired under warranty. Examine the warranty if possible. Has the project ever been altered? How did this affect existing equipment and structures?
2. Find examples of equipment or structures at your school and home that did and did not provide the expected payback. Which ones lost efficiency because of servicing required?

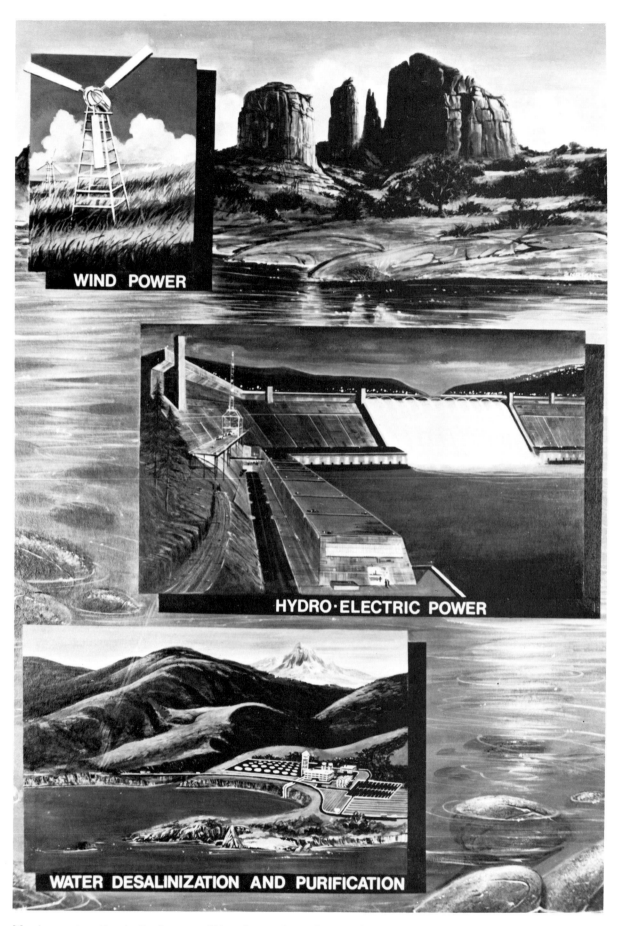

WIND POWER

HYDRO·ELECTRIC POWER

WATER DESALINIZATION AND PURIFICATION

Much construction in the future will involve projects for supplying energy and fresh water. (United Energy Corp.)

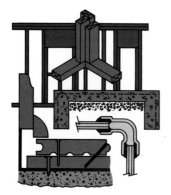

Chapter 30

CONSTRUCTION IN THE FUTURE

After studying this chapter, you will be able to:
☐ Describe how construction technology answered needs of people in the past.
☐ Discuss new problems caused by construction.
☐ List materials and resources we must conserve.
☐ Discuss ways to conserve land for housing and offices.
☐ Describe manufactured housing and projects partly built in factories.
☐ Describe new sources of energy.
☐ List ways to conserve and find new uses for land and highway building materials.
☐ Explain how knowledge of repairing can help you plan for the future.
☐ Discuss how future construction decisions depend on whether a project is energy efficient.

This chapter provides examples of how design problems were solved in the past, and what structures were built as a result. The chapter shows that the same design skills will be needed in the future. Ways to meet needs in the future will be described.

HOW PROBLEMS WERE ONCE SOLVED

The challenge for construction has always been the same. We want to improve the *quality of life* for people. Following recent developments, we also want to conserve resources.

In the past, construction has helped our nation grow. It will continue to help in the future. Construction helps when it provides the structures we need. See Figs. 30-1 and 30-2. Today we are trying to find ways to conserve (save) our resources. Construction technology can help there, too.

Provided What We Needed

After needs were identified, construction technology provided some solutions. Here are some examples of how needs were met.

Note: this chapter is too short to include more than some selected examples. Also, note that both the good and bad effects of construction technology are often given.

Housing

When cities became densely populated (meaning the number of people), we needed to save space in the inner cities. Apartment buildings as tall as 20 stories were built. These structures used land wisely and saved building materials. The result was low-cost housing.

However, some of this low-cost housing was too compact for comfortable living. The high density of people may have encouraged crime and vandalism (damage done out of boredom).

Transportation

A problem in any age is to provide fast, safe transportation. This means more than just highways. One mile of expressway takes 28 acres of land. An interchange uses about 80 acres. Some modern forms of transportation were needed to conserve land, and more transportation systems are needed.

Mass transit systems move people faster on less land. See Fig. 30-3. Mass transportation systems that run underground use very little surface land.

Water

A basic need of all people is fresh water. The demand for water has been growing while the supply of clean water has been shrinking. Cities have

Fig. 30-1. A lightweight structure was needed to enclose a large space. Many triangles of aluminum and acrylic solved the problem. (TEMCOR)

Fig. 30-2. The designer wanted to avoid supporting the bridge along its entire length with columns. Columns would be costly and would ruin the view. (U.S. Department of Transportation, Federal Highway Administration)

Fig. 30-3. The community needed to conserve land. Mass transit uses less land than roads. The system can be built over the land or under it. (HNTB Engineers)

responded by bringing more water longer distances, Fig. 30-4. Communities will need to emphasize conservation.

Agricultural land

Agricultural land is used to grow food for people and animals. Land is also used to grow crops for raw materials like cotton, lumber, and soybeans (for soy oil). We can grow more food per acre of land than ever before. But, more and more land is used for highways, housing, and other construction projects. See Fig. 30-5. Even aqueducts that bring needed water use up available land.

Communication

We can see events happening anywhere on the earth. Any telephone in America can reach over 90 percent of the telephones in the world. Communication will become even easier, faster, and cheaper in the future.

We will be building bigger and more complex structures in space. Structures like the one found in Fig. 30-6 will be more common. We cannot even imagine how many of the structures will look or work in the future.

REASONS FOR CONSERVATION

Projects can be built to conserve the resources we have. *Conserve* means to use wisely and not to waste. Our country has many resources. Among them are energy, land, minerals, and wildlife.

Fig. 30-5. A lot of cropland was used to construct these highways and the aqueduct. (U.S. Department of Transportation, Federal Highway Administration)

Fig. 30-6. Satellites have increased the speed and expanded the range of our communication.

Fig. 30-4. Cities have tried to prevent water shortages by finding more water. In the future, conservation may be more important than increasing the supply. (California Department of Water Resources)

Fig. 30-7. This huge earthen dam will store water, help control floods, and provide recreation. It will also create a reservoir that will flood thousands of acres of land. Some people think its benefits outweigh its cost. Others don't agree. Sometimes it is very hard to decide what is right. (California Department of Water Resources)

Construction can disturb the environment. The *environment* is the whole system of plants, animals, and resources that we live in. Dams can kill fish and ruin rivers. Highways that are cut into hills and forests can destroy their beauty. Is the dam shown in Fig. 30-7 worth the cost?

Structures can be wasteful of energy. They may be poorly insulated or kept too warm or too cool. Many structures are very beautiful, but wasteful.

There is a limit to the amount of land we have, so we must learn to conserve land as well. A suburb with houses 30 to 100 ft. apart uses a lot of land. When houses are spread out, more land is needed for roads and walkways. Sewer lines and electricity wires must be longer. Each dwelling has land that is seldom used.

There are many valuable minerals in the land. Minerals include various metals and rocks. Minerals are exhaustible resources. *Exhaustible resources* are resources that once they are used up there will be no more. Designers of structures need to use *inexhaustible resources* (resources that will never run out) and *renewable resources* (resources that can be grown and harvested) when they can. Wood is a renewable resource. The sun is an inexhaustible resource. *Recyclable resources* (reuseable) also conserve minerals.

HOW CONSTRUCTION TECHNOLOGY IS MEETING THE CHALLENGE

Architects, engineers, and owners are more aware of the challenge than in the past. They are building projects that provide what we need and conserve what we have.

Housing, Offices, and Transportation

A concept called *cluster housing* places house modules close together. *Modular homes* are shown in Fig. 30-8.

Some public housing projects (high-rise apartments) built in the past were too isolated from other neighborhoods. Today, public housing projects are placed on scatter sites. *Scatter sites* are sites that are mixed in with private housing. The units are attractive and people are proud to live in them.

Some homes are built in factories. This is called *manufactured housing*. These houses can be less costly than many homes built conventionally. Ma-

Fig. 30-8. There are 96 modular housing units located on this site. They were built in a plant under controlled conditions. Normal housing additions would use over five times as much land. Notice the tall crane being used to lift a module. (©Cardinal Industries, Inc.)

Fig. 30-9. These sectional housing units are built to exact standards. Materials and workers are sheltered from the weather. Most electrical and plumbing fixtures are installed at the factory. (©Cardinal Industries, Inc.)

terials are kept dry. Line production methods are used, Fig. 30-9. All of the organized work methods can lead to a higher quality unit.

Buildings can be built taller. The wind sways tall buildings. Computers, now, can sense the forces. Weights can be shifted in the structure to keep the building from swaying.

People and freight will be moved in new ways in the future. Oil and gas tankers are getting larger. They will require deeper water. More and more oil and gas terminals are needed offshore. An offshore natural gas platform that was partly built in a factory is shown in Fig. 30-10.

Energy

Energy will come from different sources. The sun will likely be a major source. Figs. 30-11 through 30-13 show a three-phase project for the future. The sun is the main source of energy for the project, but it is still important to recover most of the energy found in organic wastes like manure.

Conservation of Energy and Resources

In the future, large energy and material recovery projects will be built. Fig. 30-14 shows a steel recovery project.

Projects to burn paper wastes and other wastes will be needed. The energy found in the paper should be recovered. It is unlikely we will eliminate paper waste entirely.

When a highway's surface is worn out, it is removed and replaced. Ways are being found to reuse the materials. Used asphalt may be screened to remove large chunks for spreading on less traveled rural roads. Materials may be recycled (made like new) and reused. Some recovery equipment is shown in Fig. 30-15 and Fig. 30-16.

Fig. 30-10. Offshore gas and oil terminals are used by large tankers. This one is a six mile long pipeline trestle to an offshore terminal. (Raymond International, Inc.)

Fig. 30-11. A large solar electric project would bring many construction jobs to an area.
(United Energy Corp.)

Fig. 30-12. Biomass facilities we have at the present will be expanded. Trades including welding and pipefitting will benefit. (United Energy Corp.)

Fig. 30-13. Industries that send processed wastes to farms for reuse will benefit mainly the transportation industry. However, the need for processing buildings and their maintenance requirements can provide construction-related work.
(United Energy Corp.)

Fig. 30-14. In the picture, materials are being recovered. Steel auto bodies have value. The energy used to make new steel is saved. Small pieces of steel are moved up by the conveyor. (Hammermill, Inc.)

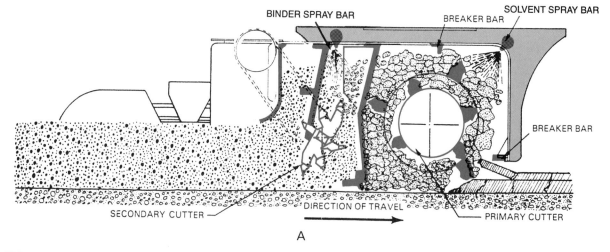

BINDER SPRAY BAR SOLVENT SPRAY BAR
BREAKER BAR

BREAKER BAR

SECONDARY CUTTER DIRECTION OF TRAVEL PRIMARY CUTTER

A

B

Fig. 30-15. Material recovery systems and machines are needed for the future. A—Cross section of a machine that breaks and collects road material, adds a solvent and binder on site, and repaves the road. B—This machine breaks up and gathers old asphalt. The used material is hauled to a nearby plant to be processed. (CMI)

Other times, the material is recycled at a separate facility, Fig. 30-16.

Land is saved by making buildings tall. Building on columns or filling-in areas to raise the grade are two other ways to use land better. See Fig. 30-17. Structures are being built over large obstacles more than ever before. Architects are also designing cities that are built under the ocean and underground. This technology can conserve agricultural land.

WHAT ROLE WILL YOU PLAY?

The future is formed by people. These people will decide what they want and then make it happen. Will you be one of those people? You can be. Begin thinking about the kind of future you want. Be sure it is good for you and the ones around you.

Fig. 30-16. Aggregate asphalt (left) and used asphalt (right) are being heated and mixed with binder and solvents. The mixture is loaded onto trucks and taken to the construction site for use. (CMI)

A

B

Fig. 30-17. New methods make it possible to build over poor land or water. A–Column supports allow the use of wet areas. (U.S. Department of Transportation, Federal Highway Administration) B–One way to conserve land is to reclaim it from the sea. (Raymond International, Inc.)

If you want the benefits a better world has to offer, you might be able to get them.

Decide what you want your future to be like. See Fig. 30-18. For example, would you choose to buy a larger house for your family if you knew how to repair gypsum board or plumbing systems? As another example, on which of the following would you like to see more work done:
- Highways?
- Railroads?
- Airports?

One project should be chosen over another if it is more efficient (uses energy wisely and saves people time).

If you imagine what your future could be like, you can begin working towards it. To put your plans into action will require some leadership. Leadership can include using advice on whether a project is good for everyone. The leader often gets ideas across through some form of public speaking. Many people working together will make the plan succeed.

ARE RAILROADS
MORE EFFICIENT
THAN HIGHWAYS?

KNOWLEDGE
OF CONSTRUCTION
TECHNOLOGY

FUTURE
CONSTRUCTION
TECHNOLOGY

ARE AIRPORTS
MORE EFFICIENT
THAN RAILROADS?

Fig. 30-18. You have many decisions to make about the future. Knowledge of construction technology can help. Future construction projects should be energy efficient.

SUMMARY

Construction technology solved some problems in the past, but it has created new problems. The challenge is to find ways to satisfy our needs, yet conserve our resources. With leadership and new ideas, construction technology can help in this task. How this is done depends on the kind of future people want.

KEY WORDS

All of the following words have been used in this chapter. Do you know their meaning?

Cluster housing
Conserve
Environment
Exhaustible resource
Inexhaustible resource
Manufactured housing
Modular home
Recyclable resource
Renewable resource
Scatter site

TEST YOUR KNOWLEDGE

Write your answers on a separate sheet of paper. Do not write in this book.

1. What are two challenges of construction in the future?
2. Construction and industry can disturb the _____ we live in.
3. Housing built in factories is called _____ housing.
4. To use resources wisely and reduce waste means to _____ resources.
5. Structures in lakes or on seashores are built on _____ or on _____.
6. List two ways construction technology can help us conserve energy.
7. How can modern construction projects conserve materials?
8. Name an exhaustible resource and an inexhaustible resource.
9. Which of the following is most needed to create a future that is good for everyone?
   a. Money sense.
   b. Hard work.
   c. Leadership.
   d. Data.
   e. Materials.
10. Name a renewable resource and a recyclable resource.

ACTIVITIES

1. Make a list of the things you can do in the next few days to conserve our resources. Find ways to save energy. Find ways to make your home, neighborhood, or school look better.
2. What plans are required to start your conservation projects?
3. Practice public speaking by introducing some conservation ideas to your class. Present arguments (reasons) to back up your ideas.

GLOSSARY

A

Access: To gain entry to a work site.

Acoustical engineer: Technical person concerned with echoes and sound transmission.

Active collector: Solar collector that uses a fluid medium for heat transfer.

Admixture: Special chemicals added to concrete. Some chemicals make the concrete set faster. Others will lengthen the setting time.

Agreement: An arrangement as to a course of action — may be verbal or written. In the case of a written agreement, it is a duly executed and legally binding contract.

Aggregate: Granular material, such as sand and gravel, commonly added to a concrete mixture.

Agricultural zone: Land designated for growing food or other materials.

Air chamber: In a plumbing system, vessel partially filled with air, which absorbs the shock of turning water off.

Air conditioner: Equipment used to remove heat and water vapor from air inside a structure. The heat is released outside the building. Water goes outside or into a drain.

Air-entrained concrete: Concrete with tiny air bubbles trapped in it.

Air exchanger: Energy-saving equipment used for preheating cold outside air being brought into a building for ventilation.

Air-supported structure: Inflatable building, made stable through air pressurization.

Alidade: A sighting instrument for topographic measuring and mapping, used in conjunction with a plane table. The upper assembly of a transit.

Appearance grades: Classifying of plywood according to quality as determined by a visual inspection of the material.

Apprentice: A person learning a trade through a program that combines on-the-job and classroom training. Trades are commonly related to skilled construction or manufacturing jobs. On-the-job training is provided by a skilled worker. Often, the classroom instruction is provided by a special teacher. In some instances, the number of classroom hours is specified by the trade.

Approval form: Document declaring official acceptance of a particular portion or a project when signed by those who made the final inspection.

Aptitude: A talent that a person may be born with or has developed.

Architect: Professional person who is skilled and knowledgeable in the design of buildings.

Architects/Engineers (A/Es): People who design the project and monitor it while it is being built. The A/Es work for the owner. The A/Es cannot change the methods used by the builder. They only make sure that the builder follows the drawings and specifications, and manage changes that are made.

Architectural drawings: That part of a set of project plans including plot plans, foundation plans, floor plans, elevations, wall sections, and details.

Asphalt: A petroleum byproduct made up of a heavy residue; used for pavements, waterproofing, shingles, etc.

Asphalt paving: Surfacing material made of aggregate and asphalt.

Assemblies: Items that need only minor assembling at the installation site.

Assembling: Manufacturing process in which parts are permanently or temporarily fastened together. There are two major ways to assemble parts into products — bonding and mechanical fastening.

Auger: A screw-like tool for boring holes, used in survey work for taking soil samples.

B

Backhoe: Machine used for excavation having a loading bucket in front that is drawn toward the machine when in operation.

Bank soil: Undisturbed earth, dense because it has little air in it.

Bar chart: Graphical presentation showing the schedule of project work completed and remaining.

Bargain collectively: A group of workers acting as one, discussing and deciding with an employer issues that relate to work.

Baseline: A definite line established in construction work from which horizontal measurements are taken when laying out a building.

Base lines: East/west lines used to establish north and south boundaries of a township.

Batt and blanket: Thermal insulation used to fill cavities between framing members.

Batter boards: Reference markings used to locate the building lines.

Bearing pile: Type of foundation used to transmit the weight of a structure to the bearing surface. Bearing piles are longer than friction piles.

Bearing surface: Place where a structure and the earth meet.

Bearing wall: An interior wall that helps to support a structure.

Bearing wall structure: Structure with usable space in it. The solid or almost solid walls hold the floors and roof above them.

Bell: Large end of cast iron pipe.

Bid: Price that is guaranteed (firm) for a period of time. Quote a price to build a project in the hopes of getting the job.

Bid bond: Form of insurance used to ensure that a bidder is able to provide specified service.

Bidding documents: Items sent out to all bidders of a project to guide them in submitting bids, includes bidding requirements, contract forms, contract conditions, specifications, drawings, and addenda. Also called a bid package.

Bidding requirements: Description of how contractors are to submit their bids, including an invitation to bidders, instructions to bidders, and all bid forms.

Bid forms: Standard forms supplied to bidders delineating items to be provided by bidder, such as costs for material, labor, overhead, and fee.

Bill of lading: A list of items shipped.

Bitumen: Thick, black, and sticky material made from oil or coal. Used in roofing.

Blasting: Demolition process that uses explosives to destroy a structure.

Boiler: Equipment used to store hot water or steam. A pump is used to force the water through the pipes to a convector or radiator.

Bond: An obligation made binding by money forfeit. Certificate of indebtedness issued by a corporation as a means of borrowing money.

Borrow pit: Centralized site serving as a source of fill material for roadbed construction. The hole remaining after construction often becomes a lake.

Brainstorming: Process in which ideas are shared by everyone so that a successful design may be developed.

Branch circuit: All of the wiring, electrical outlets, and lights controlled by one fuse or circuit breaker.

Breaking: Demolition process whereby concrete and rocks are reduced to smaller pieces with air hammers.

Brown coat: Second coat of plaster. It builds up the thickness and levels out the first coat.

Budget: An estimate of revenues and expenses.

Building code: Standards or regulations regarding the safe construction of a structure.

Building construction: Type of construction project that involves the erection of buildings (houses or skyscrapers, for example).

Building contractor: General contractor who specializes in the construction of buildings.

Building felt: Sheet of fiber soaked with asphalt.

Building permit: Form issued by state or local governments, which allows construction to begin.

Burner: Section of a convertor designed to mix the fuel and air.

C

Cap: Concrete and steel structure that connects the tops of friction piles. The cap is used to evenly spread the weight of the structure over all the piles.

Cash flow: Money coming in versus money going out of a business.

Caisson: Shaft that is driven into the earth for foundation purposes. Common materials include precast concrete, cast-in-place concrete (steel encased), and wood. Also called a pile.

Carpeting: A heavy woven fabric that is used as a floor covering.

Casting: Process, commonly used for metal and ceramic materials, in which an industrial material is first made into a liquid. The liquid material is poured or forced into a prepared mold. Then, the material is allowed or caused to solidify (become hard). The solid material is finally extracted (removed) from the mold.

Caulking: Resilient material used to seal joints and cracks in a structure. It helps to prevent rot, crumbling, and peeling paint. Caulk is used to seal cracks around doors, windows, and fixtures.

Ceiling: Overhead surface in a room.

Ceiling joist: Ceiling component that provides a nailing surface for interior ceiling materials.

Ceiling tile: Small piece of ceiling material, usually 12 or 24 inch squares.

Cemented joint: Junction point of two pieces of piping held together by a type of cement, used with plastic piping.

Cement paste: One of two parts of concrete, consisting of portland cement and water.

Center baseline: A baseline established such that it will pass through the center (plan view) of the building.

Centerline: A definite line established in construction work from which horizontal and vertical measurements are taken when laying out a pipeline, tunnel, shaft, etc.

Central air conditioning: Air conditioning serving an entire building.

Central processing unit (CPU): Part of a computer that processes information.

Certificate: A document containing a certified statement as to the truth of something.

Certificate of substantial completion: Document issued when a construction project is nearly complete.

Certificate of completion: Document declaring acceptance of a project; issued after final inspection, after all defects have been corrected and rework completed, and after the outstanding items have been reinspected.

Certification: The process of testing a worker's knowledge and skills to determine if that worker is up to some set standard.

Chair: Device used to support steel reinforcing rods in horizontal concrete structures.

Chute: A shaft in a building that is used for conveying items (laundry, mail, garbage, etc.) to lower levels using gravity.

Circuit breaker: Device inside an electrical service panel that makes the final connection between the branch circuit and the electric power coming into the panel.

Civil drawings: Working drawings used to describe the construction site before building begins.

Claim: A demand for something believed due, such as extra work to fulfill an agreement.

Clay masonry: Structural products made of clay, such as clay bricks and field tile.

Cleaning drop: Teed and capped connection off a gas piping system into which dirt settles; used for removing dirt from the gas.

Cleanout: Teed and capped connection off a piping system; uncapped and used when needed to clear blockages from the system.

Cleanup: The final building process, wherein windows, walls, and floors are washed.

Clerk: Person who takes care of receiving reports, payroll, time cards, and the majority of the other paperwork in the field office.

Client: Person who pays for a design service. May be an owner, represent an owner, or is a civic leader.

Closer unit: The last unit to be placed in a single layer of masonry units (bricks, blocks, etc.).

Cluster housing: Houses within a neighborhood placed relatively close together.

Coaxial cable: A transmission line in which one conductor completely surrounds and is insulated from a center conductor.

Coffer dam: Barrier used to keep water out of holes, made of steel sheathing.

Collar beam: Horizontal tie beam, in a roof truss, connecting two opposite rafters.

Column: A vertical structural member.

Combustion chamber: Area in a convertor where air enters, fuel burns, and fumes are exhausted (go out).

Commercial wall: Curtain wall made of standard parts.

Commercial zone: Land designated for buying and selling of products and services.

Communication systems: Processes that transmit information with electrical impulses.

Compacted soil: Earth that has been compressed to drive out most of the air.

Competitive bidding: Bidding process in which two or more contractors offer separately to build a project. Each contractor tries to get the job by bidding the lowest and finishing the project the quickest.

Composite glass: Glass, such as insulating glass and laminated glass, made of two or more layers.

Composite panels: Panels that have more than one layer.

Compressor: A machine used for increasing the pressure of a gas or vapor; used in air conditioning units.

Computer system: Computing machine consisting of a CPU, memory, input and output devices, and all other related hardware and programs (operating instructions).

Concrete: A mixture of cement paste and aggregate used in construction for footings, foundations, walls, roadways, etc.

Concrete board: Sheet of 1/2 inch thick concrete, which is reinforced with a fiberglass mesh on both sides to make it strong. Concrete board is used in damp places such as shower stalls and locker rooms.

Concrete masonry: Structural products made of concrete, such as concrete block.

Condemnation proceeding: Process of taking over private property for public purpose when an owner will not sell or asks a price that is too high. Proceeding is held in court; the court sets a fair price for the land.

Conduction: Process in which heat travels point-to-point through solid matter.

Conductor: Wire used to carry electrical power. Copper conductors are used the most. A larger conductor will carry more power than a thinner conductor.

Conduit: Tube through which insulated electric wires are run.

Conserve: Avoid wasteful use of.

Construction: The process of using manufactured goods and industrial materials to build structures on a site.

Construction documents: Working drawings and specifications for any given construction project.

Construction methods: Concerns the use of tools and techniques to build something on site.

Construction standards: Criteria defining acceptable materials and equipment for use on a construction project and acceptable methods of performing the work.

Construction technology: The study of how houses, commercial buildings, and roads are built.

Constructor: A design/build firm.

Consultant: Firm with qualified people who make recommendations on how to design, finance, manage, and promote a project.

Contract: An agreement between two or more parties.

Contract closing: Conclusion of an agreement when final payments are made in exchange for acceptance of project or possession of property.

Contractor: One that contracts to perform services.

Controlling: Comparing the results of human effort to the company goals.

Control system: Method for regulating machines or equipment automatically.

Convection: Process in which heat is transferred by contacting and entering liquids and gases.

Convector: Enclosed heating unit used to transfer heat by the convection process. Some convectors have fans to help move the air.

Converter: Used to change energy for use in heating or cooling.

Conveyance system: Method for moving people and items through a building. Examples include conveyors, escalators, elevators, etc.

Conveyor: Equipment used to tranfer material locally from one place to another.

Cost/benefit analysis: A detailed cost comparison of different alternatives that takes sales and expenses into account, producing a rate of return on investment.

Cost plus fixed fee contract: Type of contract in which materials, labor, and equipment are reimbursed to the contractor at cost. Contractor earns a fixed fee for his work that is agreed upon at the outset of the project.

Cost plus percentage of cost contract: Type of contract in which materials, labor, and equipment are reimbursed to the contractor at cost. Contractor earns a fee that is equal to some percentage of the cost. The percentage that will be used is agreed upon at the outset of the project.

Counterweight: A weight used to reduce or counter the effect of an opposing weight.

Coupler: Threaded fitting used to connect pieces of rigid conduit.

Critical path: The series of events in a construction project that take the longest time to complete. This path controls the length of time needed to complete the project.

Critical Path Method (CPM): Graphic means of showing the sequence of each task in a construction project.

Cultivating: Preparing soil for planting by breaking up and loosening.

Curb valve: Mechanical device installed in a piping system, located just off the water main between the curb and the sidewalk; used to turn water to the structure on and off.

Curing: Subjecting to a process that brings about a chemical change, such as the setting up of concrete.

Curtain wall: Frame enclosure designed for appearance and weather protection. Wall does not carry roof, floor, or any other loads.

Custom builder: Contractor who builds to the specifications of an owner.

Custom wall: Curtain wall designed from custom parts for one project.

Cut and cover: Method of building a tunnel, consisting of: digging a trench; building the tunnel structure; adding utilities and waterproofing; covering the structure.

Cut and fill: Leveling process of cutting away high spots and filling low spots.

D

Darbying: Process of smoothing wet concrete.

Deck: Roof surface laid atop framing members and designed to support the roofing material.

Decorating: In finish work, covering a surface with something other than paint.

Deed: Legal document showing ownership of land and structures thereon; property title.

Dehumidifier: Equipment that draws moist air over cool coils to remove humidity (moisture) from a space. Water condenses on the coils and runs off into a tray or into a drain.

Dehumidifying: Removing of water vapor from air.

Depreciation: The loss in resale value of equipment as it gets older.

Design development: The phase of a design process wherein ideas are refined and the best design is selected.

Detail drawing: Working drawing showing a portion of a structure on a greatly enlarged scale as it would otherwise be too small.

Detail section drawing: Section drawing showing a portion of a structure, allowing more detail to be presented.

Diagonal brace: Angled wood or metal brace used to strengthen walls.

Differential leveling: A surveying process that uses the horizontal line of sight as a reference. A vertical rod with a graduated standard is used to determine elevation of distances remote from each other.

Digester: Waste disposal system that uses bacteria to convert organic waste into methane gas.

Dimension lumber: Lumber cut to standard sizes or to sizes ordered.

Dimmer switch: Type of switch that operates by varying the voltage in a circuit.

Distribution station: An electrical facility used to supply usable power to utility customers through transformers and a network of transmission lines.

Distribution system: System that moves heated or cooled air or water throughout a building. Heat can be moved in four ways—pipes, ducts, wires, and gravity.

Divisions: The major categories detailing the scope of work in job specifications.

Down payment: The portion of a selling price paid at the time of purchase.

Downward progression: Demotion of job responsibilities.

Drainage system: Subsurface system used to remove the water from the soil around the building. Clay tile and plastic tubing are commonly used for drainage.

Dredging: Removing from water, soil that has settled.

Drilling: In earthwork, cutting a round hole deep into the earth to gather raw materials.

Drip edge: Barrier to protect the edge of decking from water that could run under it.

Driveway: A road extension giving access of vehicles to a structure.

Ductwork: Distribution system used to convey conditioned air.

Duct: Plastic, thin sheet metal, or fiber tubes used to distribute air through a building.

Dumbwaiter: A small, manually operated freight elevator. A dumbwaiter can carry more weight than a conveyor, but it is not automatic.

DWV: Common term for waste piping. Letters stand for Drain, Waste, and Vent.

E

Earnings: A reward of working that comes in the form of cash and fringe benefits.

Earthwork: Changes made in the earth when preparing a site for construction.

Easement: A strip of land, usually along a property line, reserved for public and private utilities.

Edging: Rounding concrete edges over to make them stronger.

Electrical drawings: That part of a set of project plans showing the layout, details, pertinent data schedules, and riser diagrams concerning electrical circuits.

Electrical insulation: Nonconducting material surrounding electrical conductors for isolation and safety purposes.

Electronic air cleaner: Device used to remove pollen and smoke from the air. Air is forced or drawn through the air cleaners. Electric charges throw the dirt against collection plates. The dirt sticks on the plates until it is washed off.

Electronic distance meter: Electronic survey instrument used for horizontal distance measurement.

Elevations: Working drawings showing the interior and exterior views of a structure as projected onto a vertical plane. These drawings appear as though you are looking straight at the structure.

Elevator: Equipment that raises and lowers people and freight in a structure.

Elevator constructor: Person who assembles and installs elevators and movable stairways. Elevator constructors will also service and repair the elevators and moving stairways.

Elevator service: Elevator provisions within a structure.

Engineer: Professional person whose area of expertise lies in an engineering discipline, such as civil, structural, mechanical, or electrical systems.

Engineering drawing: In manufacturing, a drawing that shows the size and shape of parts and products. Information needed to assemble the product is also given. In construction, a drawing used to indicate utilities, roads, pipelines, and other long, narrow projects.

Entrepreneur: Person who organizes, manages, and assumes the risks of a business enterprise.

Environment: Surroundings: especially, the whole system of plants, animals, and resources that are in our local area.

Environmental impact report: States whether construction will negatively affect endangered species of plants or animals.

Equipment: Items, such as water heaters or furnaces, which are often hidden or placed in a mechanical room.

Equipment rate: Cost of equipment in dollars per hour, which when multiplied by total hours gives a total equipment cost; used in estimating.

Estimate: Amount of money a person thinks it will take to build a construction project.

Estimator: A person trained to estimate costs associated with a construction project.

Evaporative cooler: Equipment used to cool and moisten air in hot, dry climates. A trickle of water keeps a loose material wet. A fan pulls outside air through the wet mat. The air is cooled and moisture is added. The cool, moist air is distributed to the enclosed space.

Excavating: To remove soil to gain space on the site, expose a bearing surface, or bury a structure.

Exchange system: System that involves people or equipment giving and getting information. Telephones, television, intercom, and computer systems are types of exchange systems.

Exhaustible resource: Material that is available only in a limited supply. Natural gas is an example of an exhaustible resource.

Extended plenum system: Duct system employing a main trunk-like plenum and smaller ducts that branch off to each register.

Exterior insulating finish system (EIFS): A watertight and airtight finish that looks like stucco.

Extruding: Forcing a semisoft solid material through a die orifice to produce a continuously formed material in some desired shape.

F

Fabric structure: Type of structure that uses air and cables to support specially treated cloth.

Fertilizer: Organic or inorganic chemicals added to soil to make plants grow better.

Feasibility study: A study done to see if a project is a good idea, the outcome of which is a clear recommendation.

Fiberglass: Glass in fibrous form used in making various products.

Fiber-optic system: Electrical system whereby signals in the form of light waves are transmitted through long, thin, flexible fibers of a transparent material.

Field engineer: Engineer assigned to the construction site to monitor progress and provide technical assistance to the workers.

Field office overhead costs: Costs that include the payroll for everyone except the workers. Surveys, office space, testing, site preparation, insurance, storage, building permits, and payments for bonds add to field office overhead costs.

Filter: A porous material used to remove dirt or other fine particles from a liquid or gas.

Finish coat: The final coat of plaster. This coat leaves a smooth, even surface.

Finish grade: A classification of lumber, reserved for more visually appealing pieces. Lumber of this grade is used where it will be seen.

Finished product: Items ready for installation that need only be set in place and connected into the system.

Finish sheen: Paint luster, including flat, satin, semigloss, etc.

Finish work: One of the final stages of construction, which includes trimming, painting, decorating, and installing.

Firefighting system: Building piping system dedicated to putting out fires.

Fish tape: Flexible steel tape used to pull wire through electrical conduit.

Fixed appliance circuit: Branch circuit devoted to one appliance.

Fixed price contract: Type of contract in which contractor is paid a fixed price for all the work performed. Also called a lump-sum contract.

Fixture: Item, such as lights, registers, and toilets, placed in homes and commercial structures.

Flange joint: Piping connection made by bolting together circular rims projecting from around each member.

Flare joint: Tubing connection wherein one end of the tubing is tapered out before joining to the other piece.

Flashing: Material used to stop leaks that occur where a roof section joins another surface. Metal, asphalt roofing, and asphalt plastic cement are used for flashing.

Float: A tool used in concrete finishing to push down large rocks and which will prepare the surface for the final finish.

Floating foundation: A special type of foundation made to carry the weight of a superstructure on unstable land.

Flood coat: Last layer of hot bitumen of a built-up roof. While it is still hot, gravel is spread over the flood coat.

Floor plans: Fully dimensioned layouts of each floor in a structure. Walls and all openings such as doors and windows are shown. Drawing appears as if you are looking down from atop the structure (with overhead floors and roof removed).

Flow test: Type of test performed on a piping system to make sure the flow is adequate and that the piping meets specifications.

Footing: The base on which the foundation rests. Footings are designed to spread the weight of the structure over a greater area.

Foundation: Substructure that extends from the bearing surface to the main structure. A foundation supports the weight of a structure. Two parts are common to all foundations—the bearing surface and the footings.

Foundation plan: Architectural drawing showing the layout of the footings and of the foundation wall of a building.

Four-way switch: A switch used in house wiring when a load is to be turned on or off at more than two places.

Framed structure: Structure with a skeleton made of wood, reinforced concrete, or steel. A skin encloses the frame.

Friction pile: Type of foundation used in weak soil. Friction piles never reach firm soil. Long poles are driven into the ground straight or at an angle.

Frost line: The depth that soil freezes in the winter. The frost line varies for different parts of the country.

Furring strip: Narrow piece of wood fastened to a wall or ceiling to provide an even nailing base for another surface.

Fuse: Protective device used to turn a circuit off when too much electricity flows. In some fuses, a small strip of lead in the fuse melts to break the connection. Fuses generally need to be replaced rather than reset.

G

Galvanizing: Zinc coating commonly used for food containers (tin cans).

Gang box: Two or more junction boxes hooked together. Somewhat large box used for storing tools, used by contractors on construction projects.

Gasket: A soft, compressible material that is squeezed at the joint. Gaskets are used to seal flange joints.

General contractor: Firm (company) that gets the overall contract for a construction project, which includes all work to be done.

General conditions: The first part of a project specifications, which sets forth rights and responsibilities existing between the owner, designer, and contractor.

General purpose circuit: Branch circuit intended for light loads; used for lights and outlets.

Geophysical method: Soil survey method that measures characteristics of soil and rock without extracting samples (resistivity and seismic methods, are two examples).

Glass block: A hollow, glass, building brick with a high insulation value.

Goal: A target of achievement, such as attaining a given job-level.

Good title: A deed free of liens.

Grievance: Formal complaint filed by a worker to express dissatisfaction of some condition, such as unfair treatment.

Ground: A low-resistance conductive pathway tied to the lowest electrical potential in a circuit.

Ground cover: Material used to cover the surface of the building site to keep the soil from washing away. Usually refers to some type of low plant, but can also refer to wood and mineral products used for mulch.

Ground fault circuit interrupter (GFCI): A fast-acting circuit breaker that trips upon sensing a current imbalance between hot and neutral wires, which means a current leak, or ground fault, is present.

Grout: A material like plaster, used to fill cracks between the floor and wall tiles. It can be made in colors to match or contrast with the tiles.

Guy wire: A wire or cable attached to something as a brace; used in landscaping to support young trees.

H

Hanger: Device used to support, or to hold in position, pipes that run horizontally.

Hardboard: Composite made of specially processed wood fibers.

Header: Structural member used to provide support over an opening such as a door or window.

Health care plan: Medical insurance policy for which an employee or union member is eligible.

Heat exchanger: Part of a convertor that takes heat from the hot exhaust medium and transfers it to cold incoming water or air.

Heat pump: System used to heat or cool a home. When it is hot inside, the heat pump works as an air conditioner; it removes heat from the inside. The heat pump works in reverse in cold weather. Heat from the outside air is removed and used to heat the inside.

Heating element: Part of an electric heater that converts electricity to heat. When electricity flows through an element, it gives off heat. Heat is removed by moving air past the element.

Heavy engineering construction: Type of construction project that involves the erection of facilities other than buildings (roadways, water control projects, piping systems, industrial complexes, electric power systems, for example).

Heavy engineering construction contractor: General contractor who specializes in heavy engineering construction projects.

Home office overhead costs: Costs that include the payroll for those who manage and work in the home office. Office rent, telephone service, utilities, advertising, and travel add to this type of overhead costs.

Horizontal conveyor: Moving sidewalk, used in some airports, for example, for moving people and small freight horizontally, not vertically.

Hot wire: An electrical lead that has an electrical potential with respect to the ground.

Humidifier: Device used to add moisture to the air.

Humidifying: Adding moisture to air.

Humidistat: Device used to control the relative humidity in an area.

Hydraulic elevator: Type of elevator that uses hydraulic power to move the car from one floor to the next; used only in low-rise buildings.

I

Incentive contract: Type of contract in which a target completion date and final cost are set up front. Contractor receives a bonus for completing ahead of schedule or under budget. Contractor is penalized if late or over budget.

Industrial construction: The construction of large industrial complexes.

Industrial union: Union that represents industrial workers (United Auto Workers, for example), as opposed to tradespeople.

Industrial wall: Curtain wall built from pieces, strips, or panels of standard material, including bricks, siding, and plywood.

Industrial zone: Land designated for manufacturing of products.

Inexhaustible resources: Resources that will never run out, such as sunlight.

Infiltration: Heat loss created by unwanted air getting into a structure. Infiltration is slowed by sealing cracks.

Infrastructure: The schools, streets, transportation and utility systems, and parks within a community or locale.

Initiating: The beginning steps involved in planning, wherein an idea is conceived and decision making gets underway.

Initiator: A person or group having an idea or seeing a need for a project.

Installing: To place something in a new structure.

Instructions to bidders: That part of a bid package setting forth directions on how to bid.

Insulation: Material that traps air; used to reduce thermal conduction. Also a protective material used to isolate wires and cables. Rubber and plastic are used to insulate wire.

Insulated glass: Said of windows having two or more layers of glass, each separated by an air barrier.

Intercom system: Exchange system, with microphone and loudspeaker, for communicating within a structure or complex.

Interest: A fee that is paid to a loaner in order to borrow money. The fee is a percentage of the amount of money borrowed.

Interior designer: Person who has a background in art rather than construction, and uses his/her skills to plan space. Interior designers combine colors, texture, and form within the space. They plan the insides of structures so they are more useful and attractive.

Interview: A formal consultation to evaluate a person's employment qualifications.

J

Invitation to bidders: A formal letter inviting selected contractors to bid on a project and describing the nature of the project. Part of a bid package.

Invoice: Bill from a vendor for payment due on purchased materials.

Jamb: Vertical sides of a doorway or window.

Job application: Form provided to a prospective employee requesting background information, such as past work, training, and skills.

Job order: A request for qualified workers to fill a certain job.

Jockey pump: Pump used in a firefighting system to maintain water pressure.

Joint compound: Pastelike material used on drywall for joints and nail dents.

Jointing: Provisions for expansion of concrete, whereby joints made in the material prevent cracking.

Joist: Horizontal support member in a structure. Flooring and ceiling materials are then attached to the joists.

Journeyman: A person who has learned a trade and works for another person; a skilled worker.

Junction box: Place where splices and connections between electrical conductors are made. A junction box provides a safe place to make these connections.

K

Keyway: Groove down the center of a footing when concrete foundation walls are to be used. The keyway strengthens the joint between the footing and the wall.

Knock-out hole: Partly punched hole in the top and bottom of a service box through which conductors are fed.

L

Labor productivity tables: Tables setting forth productivity factors in different locales across the country; used in estimating.

Labor rate: Cost of labor in dollars per hour, which when multiplied by a unit rate and a total unit quantity, or total hours, gives total cost of labor; used in estimating.

Labor union: An organization of workers formed for the purpose of advancing its members interests in respect to wages, benefits, working conditions, etc.

Lacquer: Material containing a polymer coating and a solvent. A lacquer dries when the solvent evaporates.

Laminated glass: Shatterproof glass, consisting of two or more layers of glass with a tough plastic sheet between them. The glass and plastic are bonded to form a single unit.

Laminated timber: Lumber that is glued together (arches, poles, and beams, for example).

Laminations: Heavy timbers produced from a series of layers of veneer or lumber. The grain of all layers run in the same direction. The member is held together by synthetic adhesives.

Landmarks: Fixed objects that mark a boundary or fix a location.

Landscape plan: Working drawing used to show placement of trees, shrubs, and fences, and the locations of other landscape features. Commonly used for complex construction sites.

Landscaping: Concerned with grading and planting and installing outside features like sidewalks.

Land survey: Delineation of land to establish boundaries for property.

Laser: Light Amplification by Stimulated Emission of Radiation. A laser changes electromagnetic radiation (energy waves) into light of a single color. It then amplifies (makes stronger) the light. This strong light produces heat when it strikes a surface. The heat will cut a workpiece.

Lateral progression: A change of jobs to one of the same job-level and and pay.

Lath: Narrow strips of rough wood nailed to rafters, joists, or studs and serving as a foundation for plaster.

Laying out: The initial step of masonry work, whereby masonry units are layed out in their planned positions in the structure.

Leach area: Outdoor area set aside to permit waste liquids of a septic system to pass into surrounding soil by percolation.

Leads: Corners of a masonry wall, built up prior to laying individual courses.

Lead time: The time it takes a firm to deliver its product or service to the site.

Legal action: Method of obtaining resolution to a dispute between two or more parties, conducted in a court of law.

Level: A surveying instrument used to measure elevation. Tool used to determine exact horizontal and vertical planes.

Lien: A charge upon property brought about when an owner owes money on a building.

Load: Any device that consumes electrical power.

Loading dock: Building facility for loading and unloading of trucks.

Loose fill: Thermal insulation that is poured or blown in place.

Loose soil: Earth that has been disturbed and is thereby uncompacted.

Lowest responsible bidder: In public projects, the low-bid contractor awarded the contract, provided the contractor is capable of building the project and has made a correct and complete bid.

M

Main breaker: Circuit breaker that will shut off all power to a building.

Management: Team responsible for guiding and directing company activities to ensure efficient operation.

Manufactured housing: Homes that are built in factories.

Marketing: Management function concerned with selling a product.

Markup: Amount added to a project to cover a contractor's overhead and profit. It is added to the cost price to determine the selling price.

Masonry veneer wall: Curtain wall of stone or brick.

Masonry wall: Frame enclosure of brick, concrete block, tile, or stone that can add to the load-bearing strength of a building.

Mass structure: Solid or nearly solid structure. A dam is an example of a mass structure.

Master plan: A list of tasks and the schedule to get them done.

Master planner: Person or entity, such as an architect or architectural firm, who sets the main theme for a large complex project and sees that all parts of the project design fit.

Mastic: Thick glue.

Material resource: A material used in construction, which may include natural, manufactured, assemblies, or finished products.

Mechanical drawings: That part of a set of project plans including plumbing and HVAC plans, sections and details. Also included are schedules of pertinent data, riser diagrams, and a utility site plan.

Mechanical subcontractor: Job subcontractor involved in HVAC and possibly plumbing work.

Mechanical elevator: Type of elevator that uses cables to move the car. They are used in high-rise buildings.

Mechanical system: Building system concerned with plumbing, heating/cooling, or other such mechanical-type systems.

Meridians: North/south lines used to establish east and west boundaries of a township.

Metal lath: Sheets of metal that are slit and drawn out to form openings on which plaster is spread.

Meter: Instrument used to measure quantity of a consumable service, such as electricity, water, or gas. It is placed outside the structure where it is easy to read.

Metes and bounds: Method of describing real estate having odd-shaped boundary.

Model: In manufacturing, models are final prototypes or mock-ups. Early models were used for the designer to check ideas. New models are made which include all design changes. In construction, a model is a small likeness of a construction project. Models make it easier to study each design feature.

Monitoring device: Equipment used to watch or track a place, machine, process, or physical condition.

Monitoring system: System that watches a place, a machine, or a process.

Monument: A natural or artificial structure representing a survey point, marking horizontal position and elevation.

Mortar: Special concrete made with cement, sand, water, and lime.

Mortgage: A loan for real estate, normally having a payback period, for the amount loaned plus interest, of 15 years or more.

Moving stairs: Steps that move in conveyor fashion to transport people between floors of a building. Also called an escalator.

Mulch: Covering of small wood chunks or other material spread over the soil or mixed with the soil. Used to prevent the soil from drying out.

Mullion: Vertical bar separating two windows in a multiple window.

N

Negotiated contract: Type of contract common to private owners in which the owner and contractor talk about the work and price.

Negotiating: Give-and-take process conducted between concerned parties toward the end of reaching a mutually agreeable decision.

Neutral: Grounded conductor from the load back to the power source, completing an electrical circuit.

Nipple: A short piece of pipe, threaded externally on both ends, used to couple piping.

Notice: Public and, sometimes, private notification of contractors to submit bids for an impending public project.

Notice of award: Document that serves as the contract until the actual contract is drawn up and signed.

Notice to proceed: Document that lets the contractor begin planning to build, and sets the starting and ending dates for the project.

O

Offset baseline: A baseline established such that it is set at some distance away from the planned perimeter of the building.

Offsetting: A method of surveying whereby sightings are taken off to the side of an actual boundary to avoid an obstacle in the line of sight.

Operating: Carrying on principle planning and practical work involving production and maintenance.

Oriented strand board: A panel made from large chips or flakes of wood that are aligned and glued together. Sometimes referred to as flakeboard.

Overhead: Business expenses not chargeable to a particular part of a project, such as rent, insurance, heat, etc.

P

Paint: A coating that changes from a liquid to a solid by means of polymerization (linking of molecules into strong chains). Many paints have a coloring agent added.

Painting: Finish work that protects the structure from wear and weather.

Paper tape: Tape placed over joint compound in drywall application.

Parging: Layer of mortar applied to the outside of concrete block walls to prevent moisture from seeping into the structure. This layer is commonly covered with a coat of asphalt.

Particleboard: A panel made from chips, shavings, or flakes of wood held together with a synthetic adhesive. The actual name of the material comes from the type of particles used. The most common types of particleboard are standard particleboard, waferboard, flakeboard, and oriented strand board.

Partition wall: Nonbearing wall that only divides a space, and does not add any support to the structure.

Passenger elevator: Equipment that moves people from one floor to the next in their homes or workplaces.

Passive collector: Solar collector that stores heat in some sort of large mass, such as a barrel of water or a concrete wall.

Patrol: The action of traversing a district for the maintenance of security.

Paving train: Succession of large paving equipment used to place concrete or asphalt roadways.

Payback: Money saved down-the-line as a result of some initial expenditure. After some period of time, money saved pays for the expenditure; thereafter, the money saved is surplus.

Payment bond: Insurance that all workers and suppliers will be paid if a contractor goes bankrupt.

Pedestal: Aboveground enclosure housing terminations for underground telephone cable.

Performance bond: Insurance that a project will be built if a contractor defaults on a contract.

Philadelphia rod: A painted pole with markings on it, used in reading elevations.

Photohead: Electronic device that detects light levels.

Pier: Type of foundation used to transmit the weight of the structure to a bearing surface. Piers are usually larger in diameter than piles.

Pigment: The color portion of paint.

Pile cap: A structural member placed on the top of a pile and used to distribute loads from the structure to the pile.

Pipefitter: Tradesperson who installs and repairs piping for industrial structures.

Placement service: Assistance provided to help people find work.

Plane table: A surveying instrument consisting of a drawing board mounted on a tripod and fitted with a compass and ruler. It is used to plot survey lines while in the field.

Planning: Thinking out the steps of a project before it begins.

Plans: Drawings depicting the work associated with a construction project.

Planting schedule: A schedule laying out the timing for the placement of new landscaping—trees, then sod, then flowers, for example.

Plaster: Mixture of sand, lime, and water, which is applied to walls and ceilings as a decorative covering.

Plat: A map showing land ownership, boundaries, and subdivisions.

Plates: Used as a nailing surface for the studs at the top and bottom. Double top plates add strength, keep walls straight, and support the ceiling and roof.

Platform construction: Type of construction commonly used for houses and smaller multistory structures. The floor joists of each story rest on the top plates of the story below. The bearing walls and partitions rest on the subfloor of each story.

Plenum: Chamber that joins several ducts to an air inlet or outlet.

Plumb bob: A weight suspended on a string, used to locate a point directly under another point.

Plumbing subcontractor: Job subcontractor involved in installing water, gas, and DWV systems.

Ply: Each layer of hot bitumen covered with building felt, which is applied as roofing.

Plywood: A panel composed of a core (middle layer), face layers of veneer (thin sheets of wood), and usually crossbands. The grain of the core and the face veneers run in the same direction. Except in thin plywood, the crossbands are at right angles to the face veneers.

Pneumatic tube: Conveying equipment used to carry messages and lightweight pieces or parts. The message or part is placed in a bullet-shaped case. The case is put into the inlet of a tube. Moving air in the tube pushes the case to its destination.

Potable water: Water that is pure enough to drink.

Power generating plant: Industrial facility that transforms nonelectric energy into electricity for consumer use.

Power system: A type of electrical system that supplies electrical energy needed to run mechanical equipment.

Predesign: The phase of a design process wherein the project is evaluated and defined, including the program, schedule, budget and financial package.

Prefinished molding: Trim member that has a surface color or coating already on it. It only needs to be cut and nailed in place.

Prehung door: Door unit consisting of the door, frame, hinges, and jamb. The entire unit is placed into the rough opening in the structure.

Preservative treatment: Process applied to wood to protect it against decay, mold, and insects.

Pressure treatment: A preservative treatment wherein pressure is applied to force chemicals deep into the wood.

Pressure-treated wood: Wood that has been subjected to pressure treatment; used in landscaping work as it lasts long and does not rot quickly.

Preventative maintenance: Minor routine upkeep of equipment to prevent major problems later.

Priority work: Work that is done before other work.

Private project: Project funded privately and, often, for the benefit of an owner.

Process fluids: Fluids, such as cooling water, chemicals, oils, and compressed air, that are used for industrial processing.

Procurement: Obtaining of material needed for a construction job.

Production: Changing the form of materials to add to their worth either on a site or in a factory. Production includes construction and manufacturing.

Productivity factor: The ratio of actual labor output to expected labor output for any given job. The ratio of how well labor performs on average in a given locale compared to some standard reference. The factor, multiplied by total labor hours, will adjust the estimate to account for productivity.

Professional: A person engaged in a job that requires specialized knowledge and, often, extensive academic preparation.

Profit: The money left over after an enterprise pays all expenses.

Program: Part of a computer system that provides instructions to the central processing unit.

Project manager: Person who leads the work on the job site. The project manager may be in charge of one large project or more than one smaller projects.

Public hearing: A public meeting to present and discuss a proposed project or law.

Public project: Project funded by local, state, or federal taxes for the purpose of benefiting the general public.

Punch list: List of defects found in a newly completed structure that must be corrected by the responsible firm.

Purchase order: An order sent to a vendor to buy material from the vendor, often written up in the company's purchasing department from a purchase requisition.

Purchase requisition: A request from an authorized person, such as a job superintendent, for acquisition of material or equipment and from which a purchase order is drawn up.

Purchasing agent: One involved in procurement of materials for a construction job.

Q

Qualified bidder: A bidding contractor who is qualified to build a structure.

Qualified builder: A contractor who is capable of building a project.

Qualified inspector: Competent person involved in giving official approval of a particular aspect of a project, and who has expertise in the subject matter.

R

Raceway: A channel used to hold and protect wires.

Radial system: Duct system employing a central plenum and ducts that radiate out from it.

Radiator: Equipment that transfers heat from hot water or steam to the air.

Radiation: Transmission of heat energy through space.

Rafter: Main part of a roof frame. Rafters give the roof its shape.

Rebar: Bars set in concrete to reinforce the concrete: reinforcement bar.

Receptacle: Device used to quickly disconnect and connect equipment to a circuit. Different kinds of receptacles are used for different voltages.

Recorder: Member of a survey party who takes notes on survey findings.

Recyclable resources: Materials that may be reprocessed and reused.

Recruiting: Finding new employees, or people who are capable of performing a certain job.

Reference: A person to whom inquiries as to character or ability of another can be made.

Reflective insulation: Thermal insulation used to retard the flow of heat by use of a material (aluminum foil, for example) to reflect heat radiation.

Register: Fixture through which conditioned air enters a room.

Regulator: A device that maintains a desired quantity at a set value.

Relative humidity: The amount of water vapor in the air. A relative humidity of 30-50 percent is preferred.

Relay station: In a telephone system, structure used to strengthen a radio wave and send it to the next station or exchange.

Release: Discharge from obligation or responsibility.

Release of claim: Relinquishment, or dropping, of a demand for something previously believed due.

Release of lien: Document given to the owner by the contractor when a project is complete stating, if the case may be, that nothing is owed against the project.

Remodeling: Reconstructing or altering the structure of a project.

Renewable resource: Biological material that can be replenished. Trees are an example of renewable resources.

Residential zone: Land designated for dwellings (houses, condominiums, etc.).

Resilient flooring: Flooring material made of plastic and fibers. Available in rolls or squares. The material gives a little when you walk on it.

Resistance wire: Heating element used in some radiant heaters.

Resistivity: In heat transfer, ability of a material to insulate. Resistivity is measured as an R-value. The higher the R-value, the better it insulates.

Resistivity method Soil survey method conducted by testing how well the soil conducts electricity. Differences in conductivity (or resistivity) indicate subsurface characteristics, such as presence of water.

Resources: The component parts making up an operation. In construction, this includes materials, equipment, workers, methods, and management.

Restoring: Bringing back to a former state.

Resurvey: Conduct a survey on land that was previously surveyed, often to check known points or get more information.

Retaining wall: A barrier erected to hold back earth.

Retirement plan: A company- or union-sponsored program of setting money aside for the time beyond one's active working life.

Reverberation characteristics: Qualities that relate to how long it takes for sound to be muffled in a room. Carpets, furniture, and people in a room reduce reverberation.

Ridge board: Horizontal board of a pitched roof to which common rafters are nailed.

Rigid board: Thermal insulation made of a stiff material, such as foam plastic.

Rigid conduit: Conduit made of metal pipes of standard weight and thickness and threaded at both ends.

Roadway: Flat structure used to support vehicles.

Rolling: Forming process that uses a rotating applied force to change the thickness of a piece of material.

Roof truss: Roofing component consisting of both rafters and ceiling joists. Commonly used for modern roof framing.

Roughing-in: Process of placing mechanical and electrical systems in floors, walls, and ceilings.

Rough opening: Any unfinished opening in a building.

Routine maintenance: Upkeep that is required from time to time.

Runoff water: Water that comes from rain and melting snow and goes into storm sewers.

S

Salvage: Save and recycle parts of a structure, dismantling it in the process.

Sanitary landfill: Solid waste disposal site where waste items are dumped and covered soon thereafter.

Sanitary sewer: Sewer used to convey waste water, including sewage water. Does not convey rainwater, surface water, and/or ground water.

Sanitary system: Used to provide clean water and to get rid of household and human waste.

Sash: A frame for window glass.

Scarifying: To break up and loosen the surface of (as earth or a road).

Scatter site: Public-housing site that is mixed in with private housing.

Scheduled maintenance: Upkeep that is performed on a regular basis.

Scheduling: Deciding when each production activity will take place and when it will be finished.

Scow: Barge that unloads from the bottom.

Scraper: Machine used to excavate, haul, or push dirt. Also called an earthmover.

Scratch coat: First coat of plaster.

Screeding: A straightedge used to level newly placed concrete.

Section drawings: Drawings showing a view of a structure whereby the structure appears to be cut in two, either horizontally or vertically.

Seeding: Method of lawn installation, whereby it is planted on site from seed.

Seismic method: Soil survey method conducted by testing how well sound travels in the soil. Reflection time of sound wave, indicated by sensitive electronic equipment, indicates subsurface characteristics, such as density of soil.

Septic system: Private sewage system that relies on natural bacterial action in an underground tank to disintegrate waste products.

Service: Maintain and repair any part of a facility not covered by warranty or after the warranty period is over.

Service cable: Electrical power conductor used to carry power from the service transformer to the structure. The size of conductors (wires) in the service cable depends on the amount of power needed.

Service elevator: Mechanical system used to move freight from one floor to the next. They are often larger than passenger elevators and move slower.

Service entrance: Used to connect the meter with the service panel.

Service mast: A pipe or conduit through which overhead cables enter a building.

Service panel: Steel electrical box that houses main and branch circuit breakers.

Service transformer: Equipment that lowers the voltage to that required in a structure.

Sewage: Refuse liquids or waste matter carried off by sewers.

Sheathing: Outer covering for a structure.

Sheave: A grooved wheel or pulley.

Shelter: A structure that offers protection from the elements.

Shim: Wood wedge, usually sawn cedar shingles, used to adjust and fill a space between two adjacent components.

Shingle: A small, thin piece of roofing material for laying in overlapping rows over the roof deck.

Shop drawings: Drawings made up by the individual contractors of a construction project, presenting that facet of the job in greater detail to assist the contractor in its part of the construction.

Shoring: Material acting as a prop to support earth and prevent cave-ins; used in excavation work for safety.

Shutoff valve: Mechanical device installed in a plumbing system, near the fixture, to shut water off for servicing the fixture.

Siamese connection: A piping unit with two caps, located on the outside of some buildings, to which two pumper fire trucks may hook up; used for fighting fires.

Sighting level: Surveying instrument for rough measurements, used to find a line of sight.

Sill: The horizontal member at the base of a window or door frame.

Single-throw switch: A switch in which only one set of contacts need be moved to open or close a circuit.

Sink hole: Depression in land surface caused when an underground cavern caves in.

Site plan: Working drawing showing the final contour and placement of a structure on a site. The site plan includes the boundary lines.

Site work: That part of a construction project related to the preparation of the jobsite for placement of structures and finish work. Examples include demolition, excavation, backfill, and paving.

Sky lobby: Transfer point where people change elevators in very tall buildings.

Slake: Become soft or crumble upon exposure to moisture.

Sleepers: Strips of wood laid over a rough concrete floor to which finished wood is nailed.

Slope stakes: In roadwork, landmarks placed along inclined terrain indicating cut and/or fill boundaries for the roadbed.

Sludge: Precipitated solid matter produced by water and sewage treatment processes.

Small appliance circuit: Branch circuit designed for kitchen use; used to supply power for toasters, mixers, etc.

Sodding: Method of lawn installation, whereby it is planted from sod brought in from a sod farm.

Solar energy: Type of energy given off by the sun. It is free and will not run out. However, it is hard to collect, convert, and distribute.

Solder: A metal or metallic alloy used when melted to join metallic surfaces.

Solid waste system: Any system designed to handle all the solid waste produced.

Sound: Radiant energy that is transmitted by pressure waves and that causes the sense of hearing.

Sound transmission: Occurs when sound moves through a barrier. Sound transmission is reduced by increasing the barrier's mass.

Specifications: Written description of standards and items that are not clearly shown in the working drawings; part of the construction documents.

Spigot: The small end of a cast iron pipe.

Spoil: Soil taken out of an excavation.

Spread footing: A wide, shallow footing usually made of reinforced concrete.

Sprinkler system: Fixture used to water grass and flowers.

Square: Measure of roofing material that covers 100 square feet.

Stadia leveling: A surveying process used to determine elevations of very hilly land. It is faster than differential leveling, but less accurate. Line of sight for this technique need not be horizontal.

Stain: Colored wood finish used to change the color of the wood without covering the grain.

Standpipe system: Manual firefighting system within a building. Water is held in a large vertical pipe, maintaining it under pressure, and is accessed in the event of fire.

Stock: Materials that do not need to be ordered as they are on hand.

Stop work notice: Notification that work is not to proceed until changes are made to a structure. Work can be stopped if there is no permit or if construction does not follow the building code.

Storm sewer: Sewer used to convey rainwater, surface water, and/or ground water.

Strength: A material's ability to bear a mechanical load. There are four strengths: tensile (tension), compressive, shear, and torsion.

Stringer: A long horizontal member used to support a floor or to connect uprights in a frame. An inclined member supporting the risers and treads of a stairway.

Structural drawings: That part of a set of project plans including plan views, elevations, and details describing the main support system for the structure.

Structural grade: A classification of lumber, reserved for longer, stronger, and straighter pieces.

Stucco: A plasterlike material that may be applied to the outside wall or other exterior surface of a structure.

Stud: Main vertical member of a frame structure.

Subcontractor: Contractor hired by another contractor to perform specialized services in association with some construction project.

Subfloor: Surface that provides weight-bearing strength for a structure. It is used to store materials and as a work surface. In large structures and basements, the subfloor is made of concrete. In wood-frame structures, it is plywood, lumber, or flakeboard.

Subgrade: The leveled surface upon which foundation is poured.

Substructure: The part of a structure that is below the surface of the ground, including pilings, footings, foundation, etc.

Subsurface survey: Survey conducted to describe the soil under the surface of the ground.

Superintendent: Person who works on the master schedule with the people in the home office. On the site, the superintendent manages the field office and leads each of the major tasks on the master schedule.

Superstructure: The part of a structure that is above ground. It is secured in place and sits on the foundation.

Supplemental conditions: Special conditions set forth in a project specification manual, below the general conditions, that are specific to the particular job.

Supplier: Firm that provides materials and equipment.

Surface survey: Survey conducted to describe land characteristics and locate boundaries and structures.

Survey: Process used to measure and describe land.

Survey party: A group of people assigned to a survey job.

Suspended ceiling: A ceiling that is hung from either the roof or a higher ceiling, used to cover parts of a buildings mechanical system.

Sweat joint: A type of connection made by the union of two pieces of copper tubing coated with solder. The pieces are pressed together and heat applied until the solder melts.

Swell: Measure of volumetric expansion of soil brought on by disturbing it and thereby increasing the void volume.

T

Tack strip: Strip of material used to hold carpet in place. Strips of plywood with short nails, or metal strips with hooks are placed around the perimeter of a room to hold the carpet edge.

Taconite: Low-grade ore commonly used in steelmaking.

Tag line: Cable connected to steel beams to keep the beams from swinging when lifting them.

Tax assessor: A person who sets the value of property after construction for tax purposes.

Telephone system: Exchange system used to transmit a person's voice.

Television system: Exchange or monitoring system used to transmit pictures and sounds.

Temperature sensor: Device that senses temperature and converts the value into a signal that is useful for a controller.

Tempered glass: Heat-treated glass that is hard to break and harmless if broken. Also called safety glass.

Temporary building: A nonpermanent structure erected at a jobsite to serve as an office, restroom, shop, etc., with the intent of being dismantled after completion.

Tensile structure: A structures supported by strong masts and heavy cable.

Terminal: Part of a computer system that is used to send and receive information.

Terms: Description of the requirements of both parties involved in a contract.

Terrazzo: Finish flooring material consisting of white portland cement, coloring, sand, and marble chips. This mixture is placed, smoothed, and left to harden. It is then ground, polished, and sealed.

Thermostat: Automatic temperature-control device designed to control the furnace, air conditioner, or a zone (part) of a building.

Thinwall conduit: A type of metallic conduit that can be bent.

Threaded fitting: Fitting having threaded ends.

Three-way switch: A switch that can connect one conductor to any one of two other conductors.

Tilt-up bearing walls: Walls made of reinforced concrete. After the concrete floor for a building is cast, forms for the walls are built on the floor. Steel is set in the forms. Concrete is placed, finished, and cured. The forms are then removed and the wall section is lifted into place. Steel rods that project out of the edges are welded together.

Tooling: Devices, such as cutters, holders, and clamps, that are required to manufacture many products.

Topographic survey: A survey that determines ground relief and location of natural and man-made features thereon.

Topsoil: The best soil for growing plants.

Toxic waste: Waste that is known to cause health problems and, perhaps, death. It cannot be handled in water treatment plants.

Trade discount: A monetary bonus of some amount for paying a bill prior to a stated length of time.

Trade union: Union to which construction workers belong (carpenters' union, for example), as opposed to an industrial union, which represents an entire industry.

Training program: Curriculum of more-or-less formal instruction, wherein one can further one's knowledge or skills as related to employment.

Transformer: Electrical device used to step up, step down, or isolate voltage.

Transit: A telescope used for survey work to measure vertical and horizontal angles.

Trap: A bend or dip in a water pipe, so arranged as to always be full of water, to prevent exhausting of offensive odors.

Trench: A long cut in the ground.

Trimming: Finish work which consists of covering cracks and sealing the structure from weather.

U

Underlayment: Covering for a plywood subfloor to provide a smooth surface for floor coverings. Smoothed plywood, particleboard, and hardboard are common materials for underlayment.

Unit price: Material or labor cost per unit of material, which when multiplied by a total quantity of material gives a total cost; used for estimating.

Unit rate: Labor hours per unit of material, which when multiplied by a labor rate gives a unit price for labor in cost per unit; used for estimating.

Upright support: Structural member used to transfer the pressure of the structure down to the footings. Walls are one type of upright support.

Upward progression: Promotion of job responsibilities.

Utilities: Services for general use, such as gas and electric, water and sewer, and telephone.

V

Vapor barrier: Material used to stop water vapor from entering a structure. A plastic sheet placed under the inside wall covering works well as a barrier. The vapor barrier must always be placed toward the warm side of the wall.

Varnish: Type of clear oil-based paint.

Vehicle: Part of paint that is the coating material. Oil and latex are the most common vehicles.

Vent: Pipe that lets sewer gas escape to outside air and provides a vacuum break.

Verbal contract: An agreement made with spoken words and, maybe, a handshake.

Video camera: Part of a television system, used to convert an image to an electrical wave.

Video recorder: A magnetic tape recorder capable of storing the video signals for a television program and feeding them back later to a television transmitter or directly to a receiver.

Video tape: Magnetic tape designed for recording the video signals of television programs.

Voids: Empty spaces between solid substances. As regarding soil, the smaller the volume of voids, the finer and denser the soil, and vice versa.

W

Waferboard: Composite in which wood chips are laid randomly.

Walk-in application: Method of recruiting whereby a help-wanted sign is posted at the site of employment.

Walkway: Access for pedestrian (people) traffic.

Wallboard: Panels for surfacing ceilings and walls.

Warranty: A document giving a guarantee that there are no defects.

Water control project: Concerned with restraining the force of water in oceans, lakes, and rivers.

Water heater: Plumbing equipment that provides heated water for a building.

Water main: Water supply pipe, usually buried in the vicinity of the street, that serves community residences.

Water meter: Instrument that measures quantity of water consumed; used for utility billing purposes.

Weatherstripping: Narrow strips of metal, plastic, or other material designed to slow the passage of air or moisture around doors and windows.

Worker: Human resource that runs equipment and uses materials to produce a product.

Working conditions: The features, positive or negative, concerning a job environment, such as short or long hours, exposure to the elements, job hazards, etc.

Working drawings: Drawings used to describe the shape, size, and placement of parts.

Wrecking: Demolition process whereby machines are used take down a structure.

Written contract: A duly authorized agreement made in writing, which is legally binding.

Y

Yard lumber: Lumber intended for general building purposes less than 5 inches in thickness.

Z

Zone control system: Temperature control system wherein a building is divided into zones; temperature in each zone is controlled off one thermostat.

Zoning ordinances: Rules designating zones and describing how structures are used, occupied, and placed on the land.

INDEX